Antimicrobial Drug Resistance

Antimicrobial Drug Resistance

Edited by
Ollie Gill

Larsen & Keller
www.larsen-keller.com

Antimicrobial Drug Resistance
Edited by Ollie Gill
ISBN: 978-1-63549-030-5 (Hardback)

© 2017 Larsen & Keller

⊟ Larsen & Keller

Published by Larsen and Keller Education,
5 Penn Plaza,
19th Floor,
New York, NY 10001, USA

Cataloging-in-Publication Data

Antimicrobial drug resistance / edited by Ollie Gill .
    p. cm.
Includes bibliographical references and index.
ISBN 978-1-63549-030-5
1. Anti-infective agents. 2. Drug resistance. 3. Drug
resistance in microorganisms. 4. Antibacterial agents.
5. Antibiotics. I. Gill, Ollie.
RM267 .A58 2017
615.329--dc23

The publisher's policy is to use permanent paper from mills that operate a sustainable forestry policy. Furthermore, the publisher ensures that the text paper and cover boards used have met acceptable environmental accreditation standards.

Printed and bound in the United States of America.

For more information regarding Larsen and Keller Education and its products, please visit the publisher's website www.larsen-keller.com

Table of Contents

Permissions

Index

Preface

This book provides comprehensive insights into the field of antimicrobial resistance. It talks in detail about the various methods and procedures of this field. Antimicrobial resistance refers to the stage when microbes evolve and become resistant to pharmaceutical drugs and medication. The resistance can take place via three ways, which are natural resistance, resistance acquired from other organisms and genetic mutation. The topics included in this book on this subject are of utmost significance and bound to provide incredible insights to readers. As this field is emerging at a rapid pace, the contents of this textbook will help the readers understand the modern concepts and applications of the subject. It aims to serve as a resource guide for students and facilitate the study of the discipline.

A detailed account of the significant topics covered in this book is provided below:

Chapter 1- Microbes develop resistance towards medicines that were previously used to kill them. The resistance they develop is known as antimicrobial resistance. The resistance developed can be for a number of reasons, it can either be random mutations or because of misuse of antibiotics. This chapter will provide an integrated understanding on antimicrobial resistance.

Chapter 2- Antimicrobials are agents that kill microorganisms. For example, we use antibiotics against bacteria. Using antimicrobials is not a practice developed in contemporary times but has a common practice over thousands of years. The following text on antimicrobials offers an insightful focus, keeping in mind the complex subject matter.

Chapter 3- Antibiotics are drugs used to treat bacteria. With the progress made in antibiotics, diseases such as tuberculosis have been nearly eradicated. Some of the types of antibiotics are clindamycin, dalbavancin, carbapenem, daptomycin etc. The following section will not only provide an overview, it will also delve into the topics related to it.

Chapter 4- Microbes are organisms which are either single celled or are multicellular. They are diverse and include bacteria, archaea and protozoa. Some of the types discussed within this chapter are methicillin-resistant Staphylococcus aureus, Salmonella, Escherichia coli, Campylobacter, Streptococcus pyogenes etc. Microbes are best understood in confluence with the major topics listed in the following text.

Chapter 5- Agar dilution is used to test the efficiency of antibiotics and the technique used to test the vulnerability of bacteria to antibiotics is known as broth microdilution. They are cost effective methods that are available for microbial eradication. This section discusses the methods of antimicrobial resistance detection and analysis in a critical manner providing key analysis to the subject matter.

Chapter 6- The effort to educate people of antimicrobials and to eradicate the overuse of antibiotics is known as antimicrobial stewardship. Some of the essential aspects of antimicrobial resistance discussed in this text are eagle effect, horizontal gene transfer, multidrug tolerance, resistance-nodulation-cell division superfamily, among others. This chapter elucidates the crucial theories and aspects related to antimicrobial resistance.

It gives me an immense pleasure to thank our entire team for their efforts. Finally in the end, I would like to thank my family and colleagues who have been a great source of inspiration and support.

Editor

Introduction to Antimicrobial Resistance

Microbes develop resistance towards medicines that were previously used to kill them. The resistance they develop is known as antimicrobial resistance. The resistance developed can be for a number of reasons, it can either be random mutations or because of misuse of antibiotics. This chapter will provide an integrated understanding on antimicrobial resistance.

Antimicrobial Resistance

Antimicrobial resistance (AMR) is when a microbe evolves to become more or fully resistant to antimicrobials which previously could treat it. This broader term also covers antibiotic resistance, which applies to bacteria and antibiotics. Resistance arises through one of three ways: natural resistance in certain types of bacteria; genetic mutation; or by one species acquiring resistance from another. Resistance can appear spontaneously due to random mutations; or more commonly following gradual buildup over time, and because of misuse of antibiotics or antimicrobials. Resistant microbes are increasingly difficult to treat, requiring alternative medications or higher doses—which may be more costly or more toxic. Microbes resistant to multiple antimicrobials are called multidrug resistant (MDR); or sometimes superbugs. Antimicrobial resistance is on the rise with millions of deaths every year. A few infections are now completely untreatable due to resistance. All classes of microbes develop resistance (fungi, antifungal resistance; viruses, antiviral resistance; protozoa, antiprotozoal resistance; bacteria, antibiotic resistance).

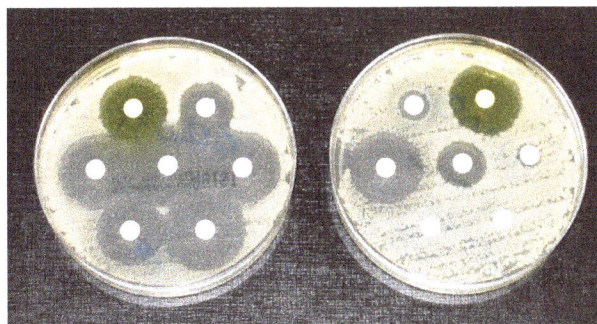

Antibiotic resistance tests: Bacteria are streaked on dishes with white antibiotic impregnated disks. Clear rings, such as those on the left show that bacteria have not grown — indicating that the bacteria are not resistant. Those on the right are fully susceptible to only three of the seven antibiotics tested.

Antibiotics should only be used when needed as prescribed by health professionals. The prescriber should closely adhere to the five rights of drug administration: the right patient, the right drug, the right dose, the right route, and the right time. Narrow-spectrum antibiotics are preferred over broad-spectrum antibiotics when possible, as effectively and accurately targeting specific organisms is less likely to cause resistance. Cultures should be taken before treatment when indicated and treatment potentially changed based on the susceptibility report. For people who take these medications at home, education about proper use is essential. Health care providers can minimize spread of resistant infections by use of proper sanitation: including handwashing and disinfecting between patients; and should encourage the same of the patient, visitors, and family members.

Rising drug resistance can be attributed to three causes use of antibiotics: in the human population; in the animal population; and spread of resistant strains between human or non-human sources. Antibiotics increase selective pressure in bacterial populations, causing vulnerable bacteria to die—this increases the percentage of resistant bacteria which continue growing. With resistance to antibiotics becoming more common there is greater need for alternative treatments. Calls for new antibiotic therapies have been issued, but new drug-development is becoming rarer. There are multiple national and international monitoring programs for drug-resistant threats. Examples of drug-resistant bacteria included in this program are: methicillin-resistant *Staphylococcus aureus* (MRSA), vancomycin-resistant *S. aureus* (VRSA), extended spectrum beta-lactamase (ESBL), vancomycin-resistant *Enterococcus* (VRE), multidrug-resistant *A. baumannii* (MRAB).

A World Health Organization (WHO) report released April 2014 stated, "this serious threat is no longer a prediction for the future, it is happening right now in every region of the world and has the potential to affect anyone, of any age, in any country. Antibiotic resistance—when bacteria change so antibiotics no longer work in people who need them to treat infections—is now a major threat to public health." Increasing public calls for global collective action to address the threat include proposals for international treaties on antimicrobial resistance. Worldwide antibiotic resistance is not fully mapped, but poorer countries with weak healthcare systems are more affected.

Definition

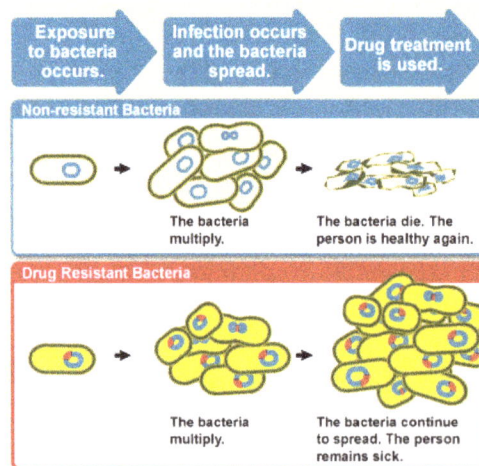

Diagram showing the difference between non-resistant bacteria and drug resistant bacteria. Non-resistant bacteria multiply, and upon drug treatment, the bacteria die. Drug resistant bacteria multiply as well, but upon drug treatment, the bacteria continue to spread.

The WHO defines antimicrobial resistance as a microorganism's resistance to an antimicrobial drug that was once able to treat an infection by that microorganism. A person cannot become resistant to antibiotics. Resistance is a property of the microbe, not a person or other organism infected by a microbe.

Causes

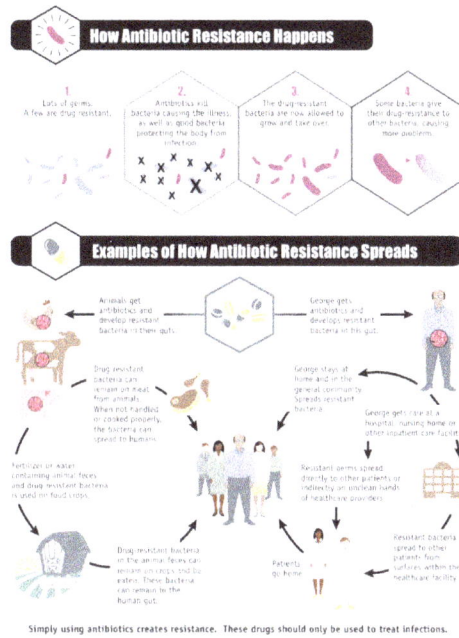

How antibiotic resistance evolves and spreads

Bacteria with resistance to antibiotics predate medical use of antibiotics by humans; however, widespread antibiotic use has made more bacteria resistant through the process of evolutionary pressure.

Reasons for the widespread use of antibiotics include:

- increasing global availability over time since the 1950s

- uncontrolled sale in many low or middle income countries, where they can be obtained over the counter without a prescription, potentially resulting in antibiotics being used when not indicated. This may result in emergence of resistance in any remaining bacteria.

Antibiotic use in livestock feed at low doses for growth promotion is an accepted practice in many industrialized countries and is known to lead to increased levels of resistance. Releasing large quantities of antibiotics into the environment during pharmaceutical manufacturing through inadequate wastewater treatment increases the risk that antibiotic-resistant strains will develop and spread. It is uncertain whether antibacterials in soaps and other products contribute to antibiotic resistance, but they are discouraged for other reasons.

Human Medicine

Increasing bacterial resistance is linked with the volume of antibiotic prescribed, as well as missing doses when taking antibiotics. Inappropriate prescribing of antibiotics has been attributed to a number of causes, including people insisting on antibiotics, physicians prescribing them as they feel they do not have time to explain why they are not necessary, and physicians not knowing when to prescribe antibiotics or being overly cautious for medical and/or legal reasons.

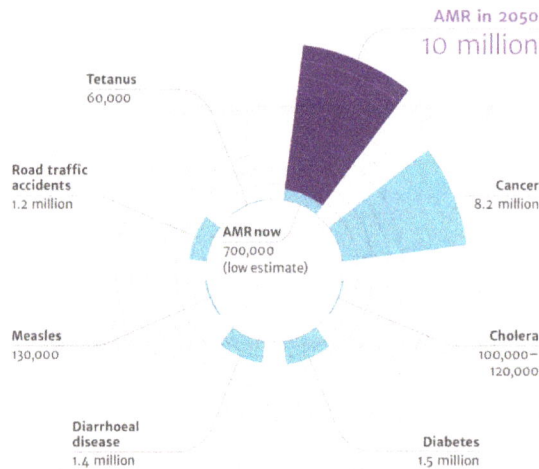

AMR in 2050
10 million

Tetanus
60,000

Road traffic
accidents
1.2 million

Cancer
8.2 million

AMR now
700,000
(low estimate)

Measles
130,000

Cholera
100,000–
120,000

Diarrhoeal
disease
1.4 million

Diabetes
1.5 million

Deaths attributable to antimicrobial resistance every year compared to other major causes of death.

Up to half of antibiotics used in humans are unnecessary and inappropriate. For example, a third of people believe that antibiotics are effective for the common cold, and the common cold is the most common reason antibiotics are prescribed even though antibiotics are useless against viruses. A single regimen of antibiotics even in compliant individuals leads to a greater risk of resistant organisms to that antibiotic in the person for a month to possibly a year.

Antibiotic resistance increases with duration of treatment; therefore, as long as an effective minimum is kept, shorter courses of antibiotics are likely to decrease rates of resistance, reduce cost, and have better outcomes with fewer complications. Short course regimens exist for community-acquired pneumonia spontaneous bacterial peritonitis, suspected lung infections in intense care wards, so-called acute abdomen, middle ear infections, sinusitis and throat infections, and penetrating gut injuries. In some situations a short course may not cure the infection as well as a long course. A BMJ editorial recommended that antibiotics can often be safely stopped 72 hours after symptoms resolve. Because individuals may feel better before the infection is eradicated, doctors must provide instructions to them so they know when it is safe to stop taking a prescription. Some researchers advocate doctors' using a very short course of antibiotics, reevaluating the patient after a few days, and stopping treatment if there are no clinical signs of infection.

Certain antibiotic classes result in resistance more than others. Increased rates of MRSA infections are seen when using glycopeptides, cephalosporins, and quinolones. Cephalosporins, and particularly quinolones and clindamycin, are more likely to produce colonisation with *Clostridium difficile*.

Factors within the intensive care unit setting such as mechanical ventilation and multiple underlying diseases also appear to contribute to bacterial resistance. Poor hand hygiene by hospital staff has been associated with the spread of resistant organisms, and an increase in hand washing compliance results in decreased rates.

Improper use of antibiotics can often be attributed to the presence of structural violence in particular regions. Socioeconomic factors such as race and poverty affect accessibility of and adherence to drug therapy. The efficacy of treatment programs for drug-resistant strains depends on whether or not programmatic improvements take into account the effects of structural violence.

Veterinary Medicine

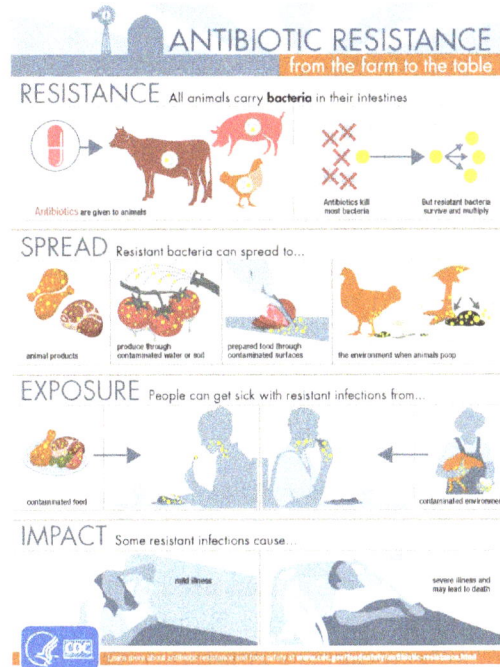

All animals carry bacteria in their intestines. Antibiotics are given to animals. Antibiotics kill most bacteria. But resistant bacteria survive and multiply.

The World Health Organization concluded that inappropriate use of antibiotics in animal husbandry is an underlying contributor to the emergence and spread of antibiotic-resistant germs, and that the use of antibiotics as growth promoters in animal feeds should be restricted. The World Organisation for Animal Health has added to the Terrestrial Animal Health Code a series of guidelines with recommendations to its members for the creation and harmonization of national antimicrobial resistance surveillance and monitoring programs, monitoring of the quantities of antibiotics used in animal husbandry, and recommendations to ensure the proper and prudent use of antibiotic substances. Another guideline is to implement methodologies that help to establish associated risk factors and assess the risk of antibiotic resistance.

United States

Eighty percent of antibiotics sold in the United States are used on livestock. The majority of these antibiotics are given to animals that are otherwise healthy. Rather, it is normal practice to mix antibiotics with livestock food to promote healthier living conditions and to encourage animal growth. The use of antibiotics in animals is to a large degree involved in the emergence of antibiotic-resistant microorganisms. Antibiotics are used in food with the intention of not only preventing, controlling, and treating diseases, but also to promote growth. Antibiotic use in animals can be classified into therapeutic, prophylactic, metaphylactic, and growth promotion uses of antibiotics. All four patterns select for bacterial resistance, since antibiotic resistance is a natural evolutionary process, but the non-therapeutic uses expose larger number of animals, and therefore of bacteria, for more extended periods, and at lower doses. They therefore greatly increase the cross-section for the evolution of resistance.

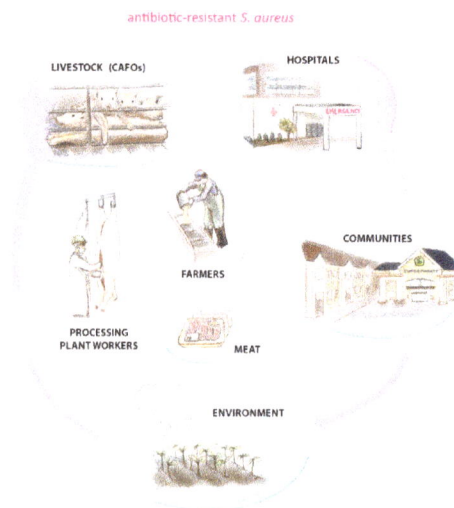

The origins of antibiotic-resistant *Staphylococcus aureus* (CAFO: concentrated animal feeding operations)

 Since the last third of the 20th century, antibiotics have been used extensively in animal husbandry. In 2013, 80% of antibiotics used in the US were used in animals and only 20% in humans; in 1997 half were used in humans and half in animals. Some antibiotics are not used and not considered significant for use in humans, because they either lack efficacy or purpose in humans, such as ionophores in ruminants, or because the drug has gone out of use in humans. Others are used in both animals and humans, including penicillin and some forms of tetracycline. Historically, regulation of antibiotic use in food animals has been limited to limiting drug residues in meat, egg, and milk products, rather than by direct concern over the development of antibiotic resistance. This mirrors the primary concerns in human medicine, where, in general, researchers and doctors were more concerned about effective but non-toxic doses of drugs rather than antibiotic resistance.

In 2001, the Union of Concerned Scientists estimated that greater than 70% of the antibiotics used in the U.S. are given to food animals (for example, chickens, pigs, and cattle), in the absence of disease. The amounts given are termed "sub-therapeutic", i.e., insufficient to combat disease. Despite no diagnosis of disease, the administration of these drugs (most of which are not significant to human medicine) results in decreased mortality and morbidity and increased growth in the animals so treated. It is theorized that sub-therapeutic dosages kills some, but not all, of the bacterial organisms in the animal — likely leaving those that are naturally antibiotic-resistant. Studies have shown, however, that, in essence, the overall population levels of bacteria are unchanged; only the mix of bacteria is affected. The actual mechanism by which sub-therapeutic antibiotic feed additives serve as growth promoters is thus unclear. Some people have speculated that animals and fowl may have sub-clinical infections, which would be cured by low levels of antibiotics in feed, thereby allowing the creatures to thrive. No convincing evidence has been advanced for this theory, and the bacterial load in an animal is essentially unchanged by use of antibiotic feed additives. The mechanism of growth promotion is therefore probably something other than "killing off the bad bugs."

Antibiotics are used in U.S. animal feed to promote animal productivity. In particular, poultry feed and water is a common route of administration of drugs, due to higher overall costs when drugs are administered by handling animals individually.

In research studies, occasional animal-to-human spread of antibiotic-resistant organisms has been demonstrated. Resistant bacteria can be transmitted from animals to humans in three ways: by consuming animal products (milk, meat, eggs, etc.), from close or direct contact with animals or other humans, or through the environment. In the first pathway, food preservation methods can help eliminate, decrease, or prevent the growth of bacteria in some food classes. Evidence for the transfer of macrolide-resistant microorganisms from animals to humans has been scant, and most evidence shows that pathogens of concern in human populations originated in humans and are maintained there, with rare cases of transference to humans.

Natural Occurrence

Naturally occurring antibiotic resistance is common. Genes for resistance to antibiotics, like antibiotics themselves, are ancient.The genes that confer resistance are known as the environmental resistome. These genes may be transferred from non-disease-causing bacteria to those that do cause disease, leading to clinically significant antibiotic resistance. In 1952 it was shown that penicillin-resistant bacteria existed before penicillin treatment; and also preexistent bacterial resistance to streptomycin. In 1962, the presence of penicillinase was detected in dormant endospores of *Bacillus licheniformis*, revived from dried soil on the roots of plants, preserved since 1689 in the British Museum. Six strains of *Clostridium*, found in the bowels of William Braine and John Hartnell (members of the Franklin Expedition) showed resistance to cefoxitin and clindamycin. Penicillinase may have emerged as a defense mechanism for bacteria in their habitats, such as the case of penicillinase-rich *Staphylococcus aureus*, living with penicillin-producing *Trichophyton*, however this may be circumstantial. Search for a penicillinase ancestor has focused on the class of proteins that must be *a priori* capable of specific combination with penicillin. The resistance to cefoxitin and clindamycin in turn was attributed to Braine's and Hartnell's contact with microorganisms that naturally produce them or random mutation in the chromosomes of *Clostridium* strains. There is evidence that heavy metals and other pollutants may select for antibiotic-resistant bacteria, generating a constant source of them in small numbers.

Environmental

Antibiotic resistance is a growing problem among humans and wildlife in terrestrial or aquatic environments. In this respect, the spread and contamination of the environment, especially through "hot spots" such as hospital wastewater and untreated urban wastewater, is a growing and serious public health problem. Antibiotics have been polluting the environment since their introduction through human waste (medication, farming), animals, and the pharmaceutical industry. Along with antibiotic waste, resistant bacteria follow, thus introducing antibiotic-resistant bacteria into the environment. As bacteria replicate quickly, the resistant bacteria that enter the environment replicate their resistance genes as they continue to divide. In addition, bacteria carrying resistance genes have the ability to spread those genes to other species via horizontal gene transfer. Therefore, even if the specific antibiotic is no longer introduced into the environment, antibiotic-resistance genes will persist through the bacteria that have since replicated without continuous exposure. Antibiotic resistance is widespread in marine vertebrates, and they may be important reservoirs of antibiotic-resistant bacteria in the marine environment.

Prevention

Mission Critical: Preventing Antibiotic Resistance (CDC report, 2014)

World Health Organization

In 2014, the WHO stated:

- People can help tackle resistance by:
    - using antibiotics only when prescribed by a doctor;
    - completing the full prescription, even if they feel better;
    - never sharing antibiotics with others or using leftover prescriptions.
- Health workers and pharmacists can help tackle resistance by:
    - enhancing infection prevention and control;
    - only prescribing and dispensing antibiotics when they are truly needed;
    - prescribing and dispensing the right antibiotic(s) to treat the illness.
- Policymakers can help tackle resistance by:
    - strengthening resistance tracking and laboratory capacity;
    - regulating and promoting appropriate use of medicines.
- Policymakers and industry can help tackle resistance by:
    - fostering innovation and research and development of new tools;
    - promoting cooperation and information sharing among all stakeholders.

Duration of Antibiotics

Antibiotic treatment duration should be based on the infection and other health problems a person may have. For many infections once a person has improved there is little evidence that stopping treatment causes more resistance. Some therefore feel that stopping early may be reasonable in some cases. Other infections; however, do require long courses regardless of whether a person feels better.

Antibiotic Usage

The Netherlands has the lowest rate of antibiotic prescribing in the OECD, at a rate of 11.4 defined daily doses (DDD) per 1,000 people per day in 2011. Germany and Sweden also have lower prescribing rates, with Sweden's rate having been declining since 2007. By contrast, Greece, France and Belgium have high prescribing rates of more than 28 DDD. It is unclear if rapid viral testing affects antibiotic use in children.

Monitoring

ResistanceOpen, an online global map of antimicrobial resistance developed by HealthMap, displays aggregated data on antimicrobial resistance from publicly available and user submitted data. The website can display data for a 25-mile radius from a location. Users may submit data from antibiograms for individual hospitals or laboratories. European data is from the EARS-Net (European Antimicrobial Resistance Surveillance Network), part of the ECDC. ResistanceMap, by the Center for Disease Dynamics, Economics & Policy, also provides data on antimicrobial resistance on a global level.

Strategies

Excessive antibiotic use has become one of the top contributors to the development of antibiotic resistance. Since the beginning of the antibiotic era, antibiotics have been used to treat a wide range of disease. Overuse of antibiotics has become the primary cause of rising levels of antibiotic resistance. The main problem is that doctors are willing to prescribe antibiotics to ill-informed individuals who believe that antibiotics can cure nearly all illnesses, including viral infections like the common cold. In an analysis of drug prescriptions, 36% of individuals with a cold or an upper respiratory infection (both viral in origin) were given prescriptions for antibiotics. These prescriptions accomplished nothing other than increasing the risk of further evolution of antibiotic resistant bacteria.

In a recent years, antimicrobial stewardship teams in hospitals have encouraged optimal use of antimicrobials. The goals of antimicrobial stewardship are to help practitioners pick the right drug at the right dose and duration of therapy while preventing misuse and minimizing the development of resistance.

There have been increasing public calls for global collective action to address the threat, including a proposal for international treaty on antimicrobial resistance. Further detail and attention is still needed in order to recognize and measure trends in resistance on the international level; the idea of a global tracking system has been suggested but implementation has yet to occur. A system of this nature would provide insight to areas of high resistance as well as information necessary for evaluation of programs and other changes made to fight or reverse antibiotic resistance.

On March 27, 2015, the White House released a comprehensive plan to address the increasing need for agencies to combat the rise of antibiotic-resistant bacteria. The Task Force for Combating Antibiotic-Resistant Bacteria developed *The National Action Plan for Combating Antibiotic-Resistant Bacteria* with the intent of providing a roadmap to guide the US in the antibiotic resistance challenge and with hopes of saving many lives. This plan outlines steps taken by the Federal government over the next five years needed in order to prevent and contain outbreaks of antibiotic-resistant infections; maintain the efficacy of antibiotics already on the market; and to help to develop future diagnostics, antibiotics, and vaccines.

The Action Plan was developed around five goals with focuses on strengthening health care, public health veterinary medicine, agriculture, food safety and research, and manufacturing. These goals, as listed by the White House, are as follows:

- Slow the Emergence of Resistant Bacteria and Prevent the Spread of Resistant Infections

- Strengthen National One-Health Surveillance Efforts to Combat Resistance

- Advance Development and use of Rapid and Innovative Diagnostic Tests for Identification and Characterization of Resistant Bacteria

- Accelerate Basic and Applied Research and Development for New Antibiotics, Other Therapeutics, and Vaccines

- Improve International Collaboration and Capacities for Antibiotic Resistance Prevention, Surveillance, Control and Antibiotic Research and Development

By following are goals set to meet by 2020:

- Establishment of antimicrobial programs within acute care hospital settings

- Reduction of inappropriate antibiotic prescription and use by at least 50% in outpatient settings and 20% inpatient settings

- Establishment of State Antibiotic Resistance (AR) Prevention Programs in all 50 states

- Elimination of the use of medically-important antibiotics for growth promotion in food-producing animals.

The World Health Organization has promoted the first World Antibiotic Awareness Week running from 16–22 November 2015. The aim of the week is to increase global awareness of antibiotic resistance. It also wants to promote the correct usage of antibiotics across all fields in order to prevent further instances of antibiotic resistance.

Vaccines

Microorganisms do not develop resistance to vaccines because a vaccine enhances the body's immune system, whereas an antibiotic operates separately from the body's normal defenses. Furthermore, if the use of vaccines increase, there is evidence that antibiotic resistant strains of pathogens will decrease; the need for antibiotics will naturally decrease as vaccines prevent infection before it occurs. However, new strains that escape immunity induced by vaccines may evolve; for example, an updated influenza vaccine is needed each year.

While theoretically promising, antistaphylococcal vaccines have shown limited efficacy, because of immunological variation between *Staphylococcus* species, and the limited duration of effectiveness of the antibodies produced. Development and testing of more effective vaccines is underway.

Alternating Therapy

Alternating therapy is a proposed method in which two or three antibiotics are taken in a rotation versus taking just one antibiotic such that bacteria resistant to one antibiotic are killed when the next antibiotic is taken. Studies have found that this method reduces the rate at which antibiotic resistant bacteria emerge in vitro relative to a single drug for the entire duration.

Development of New Drugs

Since the discovery of antibiotics, research and development (R&D) efforts have provided new drugs in time to treat bacteria that became resistant to older antibiotics, but in the 2000s there has been concern that development has slowed enough that seriously ill people may run out of treatment options. Another concern is that doctors may become reluctant to perform routine surgeries due to the increased risk of harmful infection. Backup treatments can have serious side-effects; for example, treatment of multi-drug-resistant tuberculosis can cause deafness or psychological disability. The potential crisis at hand is the result of a marked decrease in industry R&D. Poor financial investment in antibiotic research has exacerbated the situation. The pharmaceutical industry has little incentive to invest in antibiotics because of the high risk and because the potential financial returns are less likely to cover the cost of development than for other pharmaceuticals. In 2011, Pfizer, one of the last major pharmaceutical companies developing new antibiotics, shut down its primary research effort, citing poor shareholder returns relative to drugs for chronic illnesses. However, small and medium-sized pharmaceutical companies are still active in antibiotic drug research.

In the United States, drug companies and the administration of President Barack Obama have been proposing changing the standards by which the FDA approves antibiotics targeted at resistant organisms. On 12 December 2013, the Antibiotic Development to Advance Patient Treatment (ADAPT) Act of 2013 was introduced in the U.S. Congress. The ADAPT Act aims to fast-track the drug development in order to combat the growing public health threat of 'superbugs'. Under this Act, the FDA can approve antibiotics and antifungals needed for life-threatening infections based on data from smaller clinical trials. The Centers for Disease Control and Prevention (CDC) will reinforce the monitoring of the use of antibiotics that treat serious and life-threatening infections and the emerging resistance, and make the data publicly available. The FDA antibiotics labeling process, 'Susceptibility Test Interpretive Criteria for Microbial Organisms' or 'breakpoints' is also streamlined to allow the most up-to-date and cutting-edge data available to healthcare professionals under the new Act.

On 18 September 2014 Obama signed an executive order to implement the recommendations proposed in a report by the President's Council of Advisors on Science and Technology (PCAST) which outlines strategies to stream-line clinical trials and speed up the R&D of new antibiotics. Among the proposals:

- Create a 'robust, standing national clinical trials network for antibiotic testing' which will promptly enroll patients once identified to be suffering from dangerous bacterial infections. The network will allow testing multiple new agents from different companies simultaneously for their safety and efficacy.

- Establish a 'Special Medical Use (SMU)' pathway for FDA to approve new antimicrobial agents for use in limited patient populations, shorten the approval timeline for new drug so patients with severe infections could benefit as quickly as possible.

- Provide economic incentives, especially for development of new classes of antibiotics, to offset the steep R&D costs which drive away the industry to develop antibiotics.

The executive order also included a $20 million prize to encourage the development of diagnostic tests to identify highly resistant bacterial infections.

The U.S. National Institutes of Health plans to fund a new research network on the issue up to $62 million from 2013 to 2019. Using authority created by the Pandemic and All Hazards Preparedness Act of 2006, the Biomedical Advanced Research and Development Authority in the U.S. Department of Health and Human Services announced that it will spend between $40 million and $200 million in funding for R&D on new antibiotic drugs under development by GlaxoSmithKline.

One major cause of antibiotic resistance is the increased pumping activity of microbial ABC transporters, which diminishes the effective drug concentration inside the microbial cell. ABC transporter inhibitors that can be used in combination with current antimicrobials are being tested in clinical trials and are available for therapeutic regimens.

Animal Use

Europe

In 1997, European Union health ministers voted to ban avoparcin and four additional antibiotics used to promote animal growth in 1999. In 2006 a ban on the use of antibiotics in European feed, with the exception of two antibiotics in poultry feeds, became effective. In Scandinavia, there is evidence that the ban has led to a lower prevalence of antibiotic resistance in (nonhazardous) animal bacterial populations. As of 2004, several European countries established a decline of antimicrobial resistance in humans through limiting the usage antimicrobials in agriculture and food industries without jeopardizing animal health or economic cost.

United States

The United States Department of Agriculture (USDA) and the Food and Drug Administration (FDA) collect data on antibiotic use in humans and in a more limited fashion in animals. The FDA first determined in 1977 that there is evidence of emergence of antibiotic-resistant bacterial strains in livestock. The long-established practice of permitting OTC sales of antibiotics (including penicillin and other drugs) to lay animal owners for administration to their own animals nonetheless continued in all states. In 2000, the FDA announced their intention to revoke approval of fluoroquinolone use in poultry production because of substantial evidence linking it to the emergence of fluoroquinolone-resistant *Campylobacter* infections in humans. Legal challenges from the food animal and pharmaceutical industries delayed the final decision to do so until 2006. Fluroquinolones have been banned from extra-label use in food animals in the USA since 2007. However, they remain widely used in companion and exotic animals.

During 2007, two federal bills (S. 549 and H.R. 962) aimed at phasing out "nontherapeutic" antibiotics in U.S. food animal production. The Senate bill, introduced by Sen. Edward "Ted" Kennedy, died. The House bill, introduced by Rep. Louise Slaughter, died after being referred to Committee.

In March 2012, the United States District Court for the Southern District of New York, ruling in an action brought by the Natural Resources Defense Council and others, ordered the FDA to revoke approvals for the use of antibiotics in livestock that violated FDA regulations. On April 11, 2012 the FDA announced a voluntary program to phase out unsupervised use of drugs as feed additives and convert approved over-the-counter uses for antibiotics to prescription use only, requiring veterinarian supervision of their use and a prescription. In December 2013, the FDA announced the commencement of these steps to phase out the use of antibiotics for the purposes of promoting livestock growth.

Growing U.S. consumer concern about using antibiotics in animal feed has led to greater availability of "antibiotic-free" animal products. For example, chicken producer Perdue removed all human antibiotics from its feed and launched products labeled "antibiotic free" under the Harvestland brand in 2007. Consumer response was positive, and in 2014 Perdue also phased out ionophores from its hatchery and began using the "antibiotic free" labels on its Harvestland, Simply Smart, and Perfect Portions products.

Mechanisms

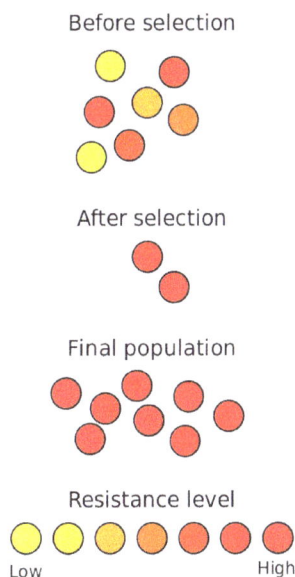

Schematic representation of how antibiotic resistance evolves via natural selection. The top section represents a population of bacteria before exposure to an antibiotic. The middle section shows the population directly after exposure, the phase in which selection took place. The last section shows the distribution of resistance in a new generation of bacteria. The legend indicates the resistance levels of individuals.

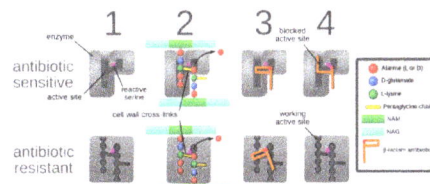

Diagram depicting antibiotic resistance through alteration of the antibiotic's target site, modeled after MRSA's resistance to penicillin. Beta-lactam antibiotics permanently inactivate PBP enzymes, which are essential for bacterial life, by permanently binding to their active sites. MRSA, however, expresses a PBP that does not allow the antibiotic into its active site.

The four main mechanisms by which microorganisms exhibit resistance to antimicrobials are:

- Drug inactivation or modification: for example, enzymatic deactivation of *penicillin* G in some penicillin-resistant bacteria through the production of β-lactamases. Most commonly, the protective enzymes produced by the bacterial cell will add an acetyl or phosphate group to a specific site on the antibiotic, which will reduce its ability to bind to the bacterial ribosomes and disrupt protein synthesis.

- Alteration of target site: for example, alteration of PBP—the binding target site of penicillins—in MRSA and other penicillin-resistant bacteria. Another protective mechanism found among bacterial species is ribosomal protection proteins. These proteins protect the bacterial cell from antibiotics that target the cell's ribosomes to inhibit protein synthesis. The mechanism involves the binding of the ribosomal protection proteins to the ribosomes of the bacterial cell, which in turn changes its conformational shape. This allows the ribosomes to continue synthesizing proteins essential to the cell while preventing antibiotics from binding to the ribosome to inhibit protein synthesis.

- Alteration of metabolic pathway: for example, some sulfonamide-resistant bacteria do not require para-aminobenzoic acid (PABA), an important precursor for the synthesis of folic acid and nucleic acids in bacteria inhibited by sulfonamides, instead, like mammalian cells, they turn to using preformed folic acid.

- Reduced drug accumulation: by decreasing drug permeability or increasing active efflux (pumping out) of the drugs across the cell surface These specialized pumps can be found within the cellular membrane of certain bacterial species and are used to pump antibiotics out of the cell before they are able to do any damage. These efflux pumps are often activated by a specific substrate associated with an antibiotic.

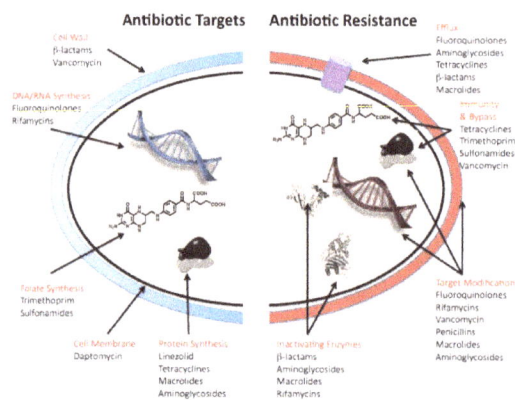

A number of mechanisms used by common antibiotics to deal with bacteria and ways by which bacteria become resistant to them.

Antibiotic resistance can be a result of horizontal gene transfer, and also of unlinked point mutations in the pathogen genome at a rate of about 1 in 10^8 per chromosomal replication. Mutations are rare but the fact that bacteria reproduce at such a high rate allows for the effect to be significant. A mutation may produce a change in the binding site of the antibiotic, which may allow the site to continue proper functioning in the presence of the antibiotic or prevent the binding of the antibiotic to the site altogether. Research has shown the bacterial protein LexA may play a key role in the acquisition of bacterial mutations giving resistance to quinolones and rifampicin. DNA damage induces the SOS gene repressor LexA to undergo autoproteolytic activity. This includes the transcription of genes encoding Pol II, Pol IV, and Pol V, which are three nonessential DNA polymerases that are required for mutation in response to DNA damage. The antibiotic action against the pathogen can be seen as an environmental pressure. Those bacteria with a mutation that allows them to survive live to reproduce. They then pass this trait to their offspring, which leads to the evolution of a fully resistant colony. Although these chromosomal mutations may seem to benefit the bacteria by providing antibiotic resistance, they also confer a cost of fitness. For example, a ribosomal mutation may protect a bacterial cell by changing the binding site of an antibiotic but it will also slow the process of protein synthesis. Additionally, a particular study specifically compared the overall fitness of antibiotic resistant strains of Escherichia coli and Salmonella typhimurium to their drug-sensitive revertants. They observed a reduced overall fitness in the antibiotic resistant strains, especially in growth rate.

There are three known mechanisms of fluoroquinolone resistance. Some types of efflux pumps can act to decrease intracellular quinolone concentration. In Gram-negative bacteria, plasmid-mediated resistance genes produce proteins that can bind to DNA gyrase, protecting it from the action of quinolones. Finally, mutations at key sites in DNA gyrase or topoisomerase IV can decrease their binding affinity to quinolones, decreasing the drug's effectiveness.

Antibiotic resistance can also be introduced artificially into a microorganism through laboratory protocols, sometimes used as a selectable marker to examine the mechanisms of gene transfer or to identify individuals that absorbed a piece of DNA that included the resistance gene and another gene of interest. A recent study demonstrated that the extent of horizontal gene transfer among *Staphylococcus* is much greater than previously expected—and encompasses genes with functions beyond antibiotic resistance and virulence, and beyond genes residing within the mobile genetic elements.

For a long time, it has been thought that, for a microorganism to become resistant to an antibiotic, it must be in a large population. However, recent findings show that there is no necessity of large populations of bacteria for the appearance of antibiotic resistance. We know now that small populations of E.coli in an antibiotic gradient can become resistant. Any heterogeneous environment with respect to nutrient and antibiotic gradients may facilitate the development of antibiotic resistance in small bacterial populations and this is also true for the human body. Researchers hypothesize that the mechanism of resistance development is based on four SNP mutations in the genome of E.coli produced by the gradient of antibiotic. These mutations confer the bacteria emergence of antibiotic resistance.

MCR-1

In November, 2015, a Chinese team first described MCR-1 gene after finding the gene in pigs and pork. It became a matter of concern when it was discovered that it could be transferred to other organisms. MCR-1 was later discovered in Malaysia, England, China, Europe, and the United States.

NDM-1

NDM-1 is an enzyme that makes bacteria resistant to a broad range of beta-lactam antibiotics.

NDM-1 was first detected in a *Klebsiella pneumoniae* isolate from a Swedish patient of Indian origin in 2008. It was later detected in bacteria in India, Pakistan, the United Kingdom, the United States, Canada and Japan.

According to A *Lancet* study, NDM-1 (New Delhi Metallo-beta-lactamase-1) originated in India, The study came to a conclusion that Indian hospitals are unsafe for treatment as Nosocomial-infections are common and with the new super-bugs on rise in India, It can be dangerous.

Organisms

Bacteria

Staphylococcus Aureus

Staphylococcus aureus (colloquially known as "Staph aureus" or a "Staph infection") is one of the major resistant pathogens. Found on the mucous membranes and the human skin of around a third of the population, it is extremely adaptable to antibiotic pressure. It was one of the earlier bacteria in which penicillin resistance was found—in 1947, just four years after the drug started being mass-produced. Methicillin was then the antibiotic of choice, but has since been replaced by oxacillin due to significant kidney toxicity. Methicillin-resistant *Staphylococcus aureus* (MRSA) was first detected in Britain in 1961, and is now "quite common" in hospitals. MRSA was responsible for 37% of fatal cases of sepsis in the UK in 1999, up from 4% in 1991. Half of all *S. aureus* infections in the US are resistant to penicillin, methicillin, tetracycline and erythromycin.

This left vancomycin as the only effective agent available at the time. However, strains with intermediate (4–8 µg/ml) levels of resistance, termed glycopeptide-intermediate *Staphylococcus aureus* (GISA) or vancomycin-intermediate *Staphylococcus aureus* (VISA), began appearing in the late 1990s. The first identified case was in Japan in 1996, and strains have since been found in hospitals in England, France and the US. The first documented strain with complete (>16 µg/ml) resistance to vancomycin, termed vancomycin-resistant *Staphylococcus aureus* (VRSA) appeared in the United States in 2002. However, in 2011, a variant of vancomycin has been tested that binds to the lactate variation and also binds well to the original target, thus reinstating potent antimicrobial activity.

A new class of antibiotics, oxazolidinones, became available in the 1990s, and the first commercially available oxazolidinone, linezolid, is comparable to vancomycin in effectiveness against MRSA. Linezolid-resistance in *S. aureus* was reported in 2001.

Community-acquired MRSA (CA-MRSA) has now emerged as an epidemic that is responsible for rapidly progressive, fatal diseases, including necrotizing pneumonia, severe sepsis, and necrotizing fasciitis. MRSA is the most frequently identified antimicrobial drug-resistant pathogen in US hospitals. The epidemiology of infections caused by MRSA is rapidly changing. In the past 10 years, infections caused by this organism have emerged in the community. The two MRSA clones in the United States most closely associated with community outbreaks, USA400 (MW2 strain, ST1 lineage) and USA300, often contain Panton-Valentine leukocidin (PVL) genes and, more frequently, have been associated with skin and soft tissue infections. Outbreaks of CA-MRSA infections have been reported in correctional facilities, among athletic teams, among military recruits, in newborn nurseries, and among men that have sex with men. CA-MRSA infections now appear endemic in many urban regions and cause most CA-*S. aureus* infections.

Streptococcus and Enterococcus

Streptococcus pyogenes (Group A *Streptococcus*: GAS) infections can usually be treated with many different antibiotics. Early treatment may reduce the risk of death from invasive group A streptococcal disease. However, even the best medical care does not prevent death in every case. For those with very severe illness, supportive care in an intensive-care unit may be needed. For persons with necrotizing fasciitis, surgery often is needed to remove damaged tissue. Strains of *S. pyogenes* resistant to macrolide antibiotics have emerged; however, all strains remain uniformly susceptible to penicillin.

Resistance of *Streptococcus pneumoniae* to penicillin and other beta-lactams is increasing worldwide. The major mechanism of resistance involves the introduction of mutations in genes encoding penicillin-binding proteins. Selective pressure is thought to play an important role, and use of beta-lactam antibiotics has been implicated as a risk factor for infection and colonization. *S. pneumoniae* is responsible for pneumonia, bacteremia, otitis media, meningitis, sinusitis, peritonitis and arthritis.

Multidrug-resistant *Enterococcus faecalis* and *Enterococcus faecium* are associated with nosocomial infections. These strains include: penicillin-resistant *Enterococcus*, vancomycin-resistant *Enterococcus*, and linezolid-resistant *Enterococcus*.

Pseudomonas Aeruginosa

Pseudomonas aeruginosa is a highly prevalent opportunistic pathogen. One of the most worrisome characteristics of *P. aeruginosa* is its low antibiotic susceptibility, which is attributable to a concerted action of multidrug efflux pumps with chromosomally encoded antibiotic resistance genes (e.g., *mexAB-oprM*, *mexXY*) and the low permeability of the bacterial cellular envelopes. *Pseudomonas aeruginosa* has the ability to produce 4-hydroxy-2-alkylquinolines (HAQs) and it has been found that HAQs have prooxidant effects, and overexpressing modestly increased susceptibility to antibiotics. The study experimented with the *Pseudomonas aeruginosa* biofilms and found that a disruption of relA and spoT genes produced an inactivation of the Stringent response (SR) in cells with nutrient limitation, which provides cells be more susceptible to antibiotics.

Clostridium Difficile

Clostridium difficile is a nosocomial pathogen that causes diarrheal disease worldwide. Diarrhea caused by *C. difficile* can be life-threatening. Infections are most frequent in people who have had recent medical and/or antibiotic treatment. *C. difficile* infections commonly occur during hospitalization.

According to a 2015 CDC report, *C. difficile* caused almost 500,000 infections in the United States over a year period. Associated with these infections were an estimated 15,000 deaths. The CDC estimates that *C. difficile* infection costs could amount to $3.8 billion over a 5-year span.

C. difficile colitis is most strongly associated with fluoroquinolones, cephalosporins, carbapenems, and clindamycin.

Some research suggests the overuse of antibiotics in the raising of livestock is contributing to outbreaks of bacterial infections such as C. difficile.

Antibiotics, especially those with a broad activity spectrum (such as clindamycin) disrupt normal intestinal flora. This can lead to an overgrowth of C. difficile, which flourishes under these conditions. Pseudomembranous colitis can follow, creating generalized inflammation of the colon and the development of "pseudomembrane", a viscous collection of inflammatory cells, fibrin, and necrotic cells. Clindamycin-resistant *C. difficile* was reported as the causative agent of large outbreaks of diarrheal disease in hospitals in New York, Arizona, Florida and Massachusetts between 1989 and 1992. Geographically dispersed outbreaks of *C. difficile* strains resistant to fluoroquinolone antibiotics, such as ciprofloxacin and levofloxacin, were also reported in North America in 2005.

Carbapenem-resistant Enterobacteriaceae

As of 2013 hard-to-treat or untreatable infections of carbapenem-resistant Enterobacteriaceae (CRE) were increasing among patients in medical facilities. CRE are resistant to nearly all available antibiotics. Almost half of hospital patients with get bloodstream CRE infections die from the infection.

Multidrug-resistant *Acinetobacter*

Acinetobacter is a gram-negative bacteria that causes pneumonia or bloodstream infections in critically ill patients. Multidrug-resistant Acinetobacter have become very resistant to antibiotics.

Drug-resistant *Campylobacter*

Campylobacter causes diarrhea (often bloody), fever, and abdominal cramps. Serious complications such as temporary paralysis can also occur. Physicians rely on ciprofloxacin and azithromycin for treating patients with severe disease although *Campylobacter* is showing resistance to these antibiotics.

Salmonella and E. coli

Infection with *Escherichia coli* and *Salmonella* can result from the consumption of contaminated food and water. Both of these bacteria are well known for causing nosocomial (hospital-linked) infections, and often, these strains found in hospitals are antibiotic resistant due to adaptations to wide spread antibiotic use. When both bacteria are spread, serious health conditions arise. Many people are hospitalized each year after becoming infected, with some dying as a result. Since 1993, some strains of *E. coli* have become resistant to multiple types of fluoroquinolone antibiotics.

Although mutation alone plays a huge role in the development of antibiotic resistance, a 2008 study found that high survival rates after exposure to antibiotics could not be accounted for by mutation alone. This study focused on the development of resistance in E. coli to three antibiotic drugs: ampicillin, tetracycline, and nalidixic acid. The researchers found that some antibiotic resistance in E. coli developed due to epigenetic inheritance rather than by direct inheritance of a mutated gene. This was further supported by data showing that reversion to antibiotic sensitivity was relatively common as well. This could only be explained by epigenetics. Epigenetics is a type of inheritance in which gene expression is altered rather than the genetic code itself. There are many modes by which this alteration of gene expression can occur, including methylation of DNA and histone modification; however, the important point is that both inheritance of random mutations and epigenetic markers can result in the expression of antibiotic resistance genes.

Resistance to polymyxins first appear in 2011. An easier way for this resistance to spread, a plasmid known as MCR-1 was discovered in 2015.

Acinetobacter Baumannii

On November 5, 2004, the Centers for Disease Control and Prevention (CDC) reported an increasing number of *Acinetobacter baumannii* bloodstream infections in patients at military medical facilities in which service members injured in the Iraq/Kuwait region during Operation Iraqi Freedom and in Afghanistan during Operation Enduring Freedom were treated. Most of these showed multidrug resistance (MRAB), with a few isolates resistant to all drugs tested.

Klebsiella Pneumoniae

Klebsiella pneumoniae carbapenemase (KPC)-producing bacteria are a group of emerging highly drug-resistant Gram-negative bacilli causing infections associated with significant morbidity and mortality whose incidence is rapidly increasing in a variety of clinical settings around the world. *Klebsiella pneumoniae* includes numerous mechanisms for antibiotic resistance, many of which are located on highly mobile genetic elements. Carbapenem antibiotics (heretofore often the treatment of last resort for resistant infections) are generally not effective against KPC-producing organisms.

Mycobacterium Tuberculosis

Tuberculosis is increasing across the globe, especially in developing countries, over the past few years. TB resistant to antibiotics is called MDR TB (Multidrug Resistant TB). Globally, MDR TB causes 150,000 deaths annually. The rise of the HIV/AIDS epidemic has contributed to this.

TB was considered one of the most prevalent diseases, and did not have a cure until the discovery of Streptomycin by Selman Waksman in 1943. However, the bacteria soon developed resistance. Since then, drugs such as isoniazid and rifampin have been used. M. tuberculosis develops resistance to drugs by spontaneous mutations in its genomes. Resistance to one drug is common, and this is why treatment is usually done with more than one drug. Extensively Drug-Resistant TB (XDR TB) is TB that is also resistant to the second line of drugs.

Resistance of *Mycobacterium tuberculosis* to isoniazid, rifampin, and other common treatments has become an increasingly relevant clinical challenge. (For more on Drug-Resistant TB, visit the Multi-drug-resistant tuberculosis page.) Evidence is lacking for whether these bacteria have plasmids. Also *M. tuberculosis* lack the opportunity to interact with other bacteria in order to share plasmids.

Neisseria Gonorrhoeae

Neisseria gonorrhoeae is a sexually transmitted pathogen that causes gonorrhea, a sexually transmitted disease that can result in discharge and inflammation at the urethra, cervix, pharynx, or rectum. It can cause pelvic pain, pain on urination, penile and vaginal discharge, as well as systemic symptoms. It can also cause severe reproductive complications. The bacteria was first identified in 1879, although some Biblical scholars believe that references to the disease can be found as early as Parshat Metzora of the Old Testament.

In the 1940s effective treatment with penicillin became available, but by the 1970s resistant strains predominated. Resistance to penicillin has developed through two mechanisms: chromasomally mediated resistance (CMRNG) and penicillinase-mediated resistance (PPNG). CMRNG involves step wise mutation of penA, which codes for the penicillin-binding protein (PBP-2); mtr, which encodes an efflux pump that removes penicillin from the cell; and penB, which encodes the bacterial cell wall porins. PPNG involves the acquisition of a plasmid-borne beta-lactamase. *N. gonorrheoea* has a high affinity for horizontal gene transfer, and as a result, the existence of any strain resistant to a given drug could spread easily across strains.

Fluoroquinolones were a useful next-line treatment until resistance was achieved through efflux pumps and mutations to the gyrA gene, which encodes DNA gyrase. Third-generation cephalosporins have been used to treat gonorrhoea since 2007, but resistant strains have emerged. As of 2010, the recommended treatment is a single 250 mg intramuscular injection of ceftriaxone, sometimes in combination with azithromycin or doxycycline. However, certain strains of *N. gonorrhoeae* can be resistant to antibiotics usually that are normally used to treat it. These include: cefixime (an oral cephalosporin), ceftriaxone (an injectable cephalosporin), azithromycin, aminoglycosides, and tetracycline.

Viruses

Specific antiviral drugs are used to treat some viral infections. These drugs prevent viruses from reproducing by inhibiting essential stages of the virus's replication cycle in infected cells. Antivirals are used to treat HIV, hepatitis B, hepatitis C, influenza, herpes viruses including varicella zoster virus, cytomegalovirus and Epstein-Barr virus. With each virus, some strains have become resistant to the administered drugs.

Resistance to HIV antivirals is problematic, and even multi-drug resistant strains have evolved. Resistant strains of the HIV virus emerge rapidly if only one antiviral drug is used. Using three or more drugs together has helped to control this problem, but new drugs are needed because of the continuing emergence of drug-resistant HIV strains.

Fungi

Infections by fungi are a cause of high morbidity and mortality in immunocompromised persons, such as those with HIV/AIDS, tuberculosis or receiving chemotherapy. The fungi candida, *Cryptococcus neoformans* and *Aspergillus fumigatus* cause most of these infections and antifungal resistance occurs in all of them. Multidrug resistance in fungi is increasing because of the widespread use of antifungal drugs to treat infections in immunocompromised individuals.

Of particular note, Fluconazole-resistant Candida species have been highlighted as a growing problem by the CDC. More than 20 Candida species of Candida can cause Candidiasis infection, the most common of which is *Candida albicans*. Candida yeasts normally inhabit the skin and mucous membranes without causing infection. However, overgrowth of Candida can lead to Candidiasis. Some Candida strains are becoming resistant to first-line and second-line antifungal agents such as azoles and echinocandins.

Parasites

The protozoan parasites that cause the diseases malaria, trypanosomiasis, toxoplasmosis, cryptosporidiosis and leishmaniasis are important human pathogens.

Malarial parasites that are resistant to the drugs that are currently available to infections are common and this has led to increased efforts to develop new drugs. Resistance to recently developed drugs such as artemisinin has also been reported. The problem of drug resistance in malaria has driven efforts to develop vaccines.

Trypanosomes are parasitic protozoa that cause African trypanosomiasis and Chagas disease (American trypanosomiasis). There are no vaccines to prevent these infections so drugs such as pentamidine and suramin, benznidazole and nifurtimox and used to treat infections. These drugs are effective but infections caused by resistant parasites have been reported.

Leishmaniasis is caused by protozoa and is an important public health problem worldwide, especially in sub-tropical and tropical countries. Drug resistance has "become a major concern".

Applications

Antibiotic resistance is an important tool for genetic engineering. By constructing a plasmid that contains an antibiotic-resistance gene as well as the gene being engineered or expressed, a researcher can ensure that, when bacteria replicate, only the copies that carry the plasmid survive. This ensures that the gene being manipulated passes along when the bacteria replicates.

In general, the most commonly used antibiotics in genetic engineering are "older" antibiotics. These include:

- ampicillin
- kanamycin
- tetracycline
- chloramphenicol

In industry, the use of antibiotic resistance is disfavored, since maintaining bacterial cultures would require feeding them large quantities of antibiotics. Instead, the use of auxotrophic bacterial strains (and function-replacement plasmids) is preferred.

Society and Culture

For the fiscal year 2016 budget, President Obama has suggested to nearly double the amount of federal funding to "combat and prevent" antibiotic resistance to more than $1.2 billion.

Since the mid-1980s pharmaceutical companies have invested in medications for cancer or chronic disease that have greater potential to make money and have "de-emphasized or dropped development of antibiotics." On January 20, 2016 at the World Economic Forum in Davos, Switzerland, more than "80 pharmaceutical and diagnostic companies" from around the world called for 'transformational commercial models' at a global level to spur research and development on antibiotics and on the "enhanced use of diagnostic tests that can rapidly identify the infecting organism."

Legal Frameworks

Some global health scholars have argued that a global, legal framework is needed to prevent and control antimicrobial resistance. For instance, binding global policies could be used to create antimicrobial use standards, regulate antibiotic marketing, and strengthen global surveillance systems. Ensuring compliance of involved parties is a challenge. Global antimicrobial resistance policies could take lessons from the environmental sector by adopting strategies that have made international environmental agreements successful in the past such as: sanctions for non-compliance, assistance for implementation, majority vote decision-making rules, an independent scientific panel, and specific commitments.

Multiple Drug Resistance

Multiple drug resistance (MDR), multidrug resistance or multiresistance is antimicrobial resistance shown by a species of microorganism to multiple antimicrobial drugs. The types most threatening to public health are MDR bacteria that resist multiple antibiotics; other types include MDR viruses, fungi, and parasites (resistant to multiple antifungal, antiviral, and antiparasitic drugs of a wide chemical variety). Recognizing different degrees of MDR, the terms extensively drug resistant (XDR) and pandrug-resistant (PDR) have been introduced. The definitions were published in 2011 in the journal *Clinical Microbiology and Infection* and are openly accessible.

Common Multidrug-resistant Organisms (MDROs)

Common multidrug-resistant organisms are usually bacteria:

- Vancomycin-Resistant Enterococci (VRE)

- Methicillin-Resistant *Staphylococcus aureus* (MRSA)

- Extended-spectrum β-lactamase (ESBLs) producing Gram-negative bacteria

- *Klebsiella pneumoniae* carbapenemase (KPC) producing Gram-negatives

- MultiDrug-Resistant gram negative rods (MDR GNR) MDRGN bacteria such as *Enterobacter species*, *E.coli*, *Klebsiella pneumoniae*, *Acinetobacter baumannii*, *Pseudomonas aeruginosa*

A group of gram-positive and gram-negative bacteria of particular recent importance have been dubbed as the ESKAPE group (*Enterococcus faecium*, *Staphylococcus aureus*, *Klebsiella pneumoniae*, *Acinetobacter baumannii*, *Pseudomonas aeruginosa* and Enterobacter species).

- Multi-drug-resistant tuberculosis

Bacterial Resistance to Antibiotics

Various microorganisms have survived for thousands of years by their ability to adapt to antimicrobial agents. They do so via spontaneous mutation or by DNA transfer. This process enables some bacteria to oppose the action of certain antibiotics, rendering the antibiotics ineffective. These microorganisms employ several mechanisms in attaining multi-drug resistance:

- No longer relying on a glycoprotein cell wall

- Enzymatic deactivation of antibiotics

- Decreased cell wall permeability to antibiotics

- Altered target sites of antibiotic

- Efflux mechanisms to remove antibiotics

- Increased mutation rate as a stress response

Many different bacteria now exhibit multi-drug resistance, including staphylococci, enterococci, gonococci, streptococci, salmonella, as well as numerous other gram-negative bacteria and *Mycobacterium tuberculosis*. Antibiotic resistant bacteria are able to transfer copies of DNA that code for a mechanism of resistance to other bacteria even distantly related to them, which then are also able to pass on the resistance genes and so generations of antibiotics resistant bacteria are produced. This process is called horizontal gene transfer.

Antifungal Resistance

Yeasts such as *Candida species* can become resistant under long term treatment with azole preparations, requiring treatment with a different drug class. Scedosporium prolificans infections are almost uniformly fatal because of their resistance to multiple antifungal agents.

Antiviral Resistance

HIV is the prime example of MDR against antivirals, as it mutates rapidly under monotherapy. Influenza virus has become increasingly MDR; first to amantadenes, then to neuraminidase inhibitors such as oseltamivir, (2008-2009: 98.5% of Influenza A tested resistant), also more commonly in immunoincompetent people Cytomegalovirus can become resistant to ganciclovir and foscarnet under treatment, especially in immunosuppressed patients. Herpes simplex virus rarely becomes resistant to acyclovir preparations, mostly in the form of cross-resistance to famciclovir and valacyclovir, usually in immunosuppressed patients.

Antiparasitic Resistance

The prime example for MDR against antiparasitic drugs is malaria. *Plasmodium vivax* has become chloroquine and sulfadoxine-pyrimethamine resistant a few decades ago, and as of 2012 artemisinin-resistant Plasmodium falciparum has emerged in western Cambodia and western Thailand. *Toxoplasma gondii* can also become resistant to artemisinin, as well as atovaquone and sulfadiazine, but is not usually MDR Antihelminthic resistance is mainly reported in the veterinary literature, for example in connection with the practice of livestock drenching and has been recent focus of FDA regulation.

Preventing the Emergence of Antimicrobial Resistance

To limit the development of antimicrobial resistance, it has been suggested to:

- Use the appropriate antimicrobial for an infection; e.g. no antibiotics for viral infections

- Identify the causative organism whenever possible

- Select an antimicrobial which targets the specific organism, rather than relying on a broad-spectrum antimicrobial

- Complete an appropriate duration of antimicrobial treatment (not too short and not too long)

- Use the correct dose for eradication; subtherapeutic dosing is associated with resistance, as demonstrated in food animals.

The medical community relies on education of its prescribers, and self-regulation in the form of appeals to voluntary antimicrobial stewardship, which at hospitals may take the form of an anti-microbial stewardship program. It has been argued that depending on the cultural context govern-ment can aid in educating the public on the importance of restrictive use of antibiotics for human clinical use, but unlike narcotics, there is no regulation of its use anywhere in the world at this time. Antibiotic use has been restricted or regulated for treating animals raised for human con-sumption with success, in Denmark for example.

Infection prevention is the most efficient strategy of prevention of an infection with a MDR or-ganism within a hospital, because there are few alternatives to antibiotics in the case of an exten-sively resistant or panresistant infection; if an infection is localized, removal or excision can be attempted (with MDR-TB the lung for example), but in the case of a systemic infection only generic measures like boosting the immune system with immunoglobulins may be possible. The use of bacteriophages (viruses which kill bacteria) has no clinical application at the present time.

References

- Swedish work on containment of antibiotic resistance – Tools, methods and experiences (PDF). Stockholm: Public Health Agency of Sweden. 2014. pp. 16–17, 121–128. ISBN 978-91-7603-011-0.

- editors, Ronald Eccles, Olaf Weber, (2009). Common cold (Online-Ausg. ed.). Basel: Birkhäuser. p. 234. ISBN 978-3-7643-9894-1.

- Rosner, Fred (1995). Medicine in the Bible and the Talmud : selections from classical Jewish sources (Augm. ed.). Hoboken, NJ: KTAV Pub. House. ISBN 0-88125-506-8.

- Ponte-Sucre, A, ed. (2009). ABC Transporters in Microorganisms. Caister Academic Press. ISBN 978-1-904455-49-3.

- Schnirring L (18 December 2015). "More MCR-1 findings lead to calls to ban ag use of colistin". University of Minnesota, Center for Infectious Disease Research and Policy. Retrieved 7 August 2016.

- Maryn McKenna (3 December 2015). "Apocalypse Pig Redux: Last-Resort Resistance in Europe". Germination (Blog). Retrieved 7 August 2016.

- Parry L (27 June 2016). "Second patient in US is infected with 'superbug' resistant to ALL antibiotics". Daily Mail. Retrieved 7 August 2016.

- President's 2016 Budget Proposes Historic Investment to Combat Antibiotic-Resistant Bacteria to Protect

Public Health The White House, Office of the Press Secretary, January 27, 2015

- Pollack, Andrew (20 January 2016). "To Fight 'Superbugs,' Drug Makers Call for Incentives to Develop Antibiotics". Davos 2016 Special Report. Davos, Switzerland: New York Times. Retrieved 24 January 2016.

- "Duration of antibiotic therapy and resistance". NPS Medicinewise. National Prescribing Service Limited trading, Australia. 13 June 2013. Retrieved 22 July 2015.

- Scales, David. "Mapping Antibiotic Resistance: Know The Germs In Your Neighborhood". WBUR. National Public Radio. Retrieved 8 December 2015.

- "FACT SHEET: Obama Administration Releases National Action Plan to Combat Antibiotic-Resistant Bacteria". whitehouse.gov. Retrieved 2015-10-30.

- Stephanie Strom (July 31, 2015). "Perdue Sharply Cuts Antibiotic Use in Chickens and Jabs at Its Rivals". The New York Times. Retrieved August 12, 2015.

Antimicrobial: An Integrated Study

Antimicrobials are agents that kill microorganisms. For example, we use antibiotics against bacteria. Using antimicrobials is not a practice developed in contemporary times but has a common practice over thousands of years. The following text on antimicrobials offers an insightful focus, keeping in mind the complex subject matter.

Antimicrobial

An antimicrobial is an agent that kills microorganisms or inhibits their growth. Antimicrobial medicines can be grouped according to the microorganisms they act primarily against. For example, antibiotics are used against bacteria and antifungals are used against fungi. They can also be classified according to their function. Agents that kill microbes are called microbicidal, while those that merely inhibit their growth are called biostatic. The use of antimicrobial medicines to treat infection is known as antimicrobial chemotherapy, while the use of antimicrobial medicines to prevent infection is known as antimicrobial prophylaxis.

The main classes of antimicrobial agents are disinfectants ("nonselective antimicrobials" such as bleach), which kill a wide range of microbes on non-living surfaces to prevent the spread of illness, antiseptics (which are applied to living tissue and help reduce infection during surgery), and antibiotics (which destroy microorganisms within the body). The term "antibiotic" originally described only those formulations derived from living organisms but is now also applied to synthetic antimicrobials, such as the sulphonamides, or fluoroquinolones. The term also used to be restricted to antibacterials (and is often used as a synonym for them by medical professionals and in medical literature), but its context has broadened to include all antimicrobials. Antibacterial agents can be further subdivided into bactericidal agents, which kill bacteria, and bacteriostatic agents, which slow down or stall bacterial growth.

Use of substances with antimicrobial properties is known to have been common practice for at least 2000 years. Ancient Egyptians and ancient Greeks used specific molds and plant extracts to treat infection. More recently, microbiologists such as Louis Pasteur and Jules Francois Joubert observed antagonism between some bacteria and discussed the merits of controlling these interactions in medicine. In 1928, Alexander Fleming became the first to discover a natural antimicrobial fungus known as *Penicillium rubens*. The substance extracted from the fungus he named penicillin and in 1942 it was successfully used to treat a *Streptococcus* infection. Penicillin also proved successful in the treatment of many other infectious diseases such as gonorrhea, strep throat and pneumonia, which were potentially fatal to patients until then.

Many antimicrobial agents exist, for use against a wide range of infectious diseases.

Chemical
Antibacterials

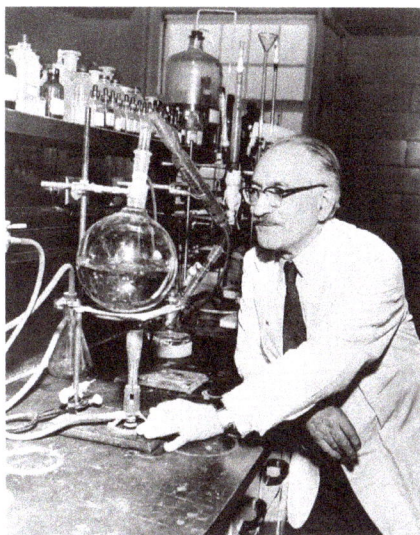

Selman Waksman, who was awarded the Nobel Prize in Medicine for developing
22 antibiotics—most notably Streptomycin.

Antibacterials are used to treat bacterial infections. The toxicity to humans and other animals from antibacterials is generally considered low. However, prolonged use of certain antibacterials can decrease the number of gut flora, which may have a negative impact on health. After prolonged antibacterial use consumption of probiotics and reasonable eating can help to replace destroyed gut flora. Stool transplants may be considered for patients who are having difficulty recovering from prolonged antibiotic treatment, as for recurrent *Clostridium difficile* infections.

The discovery, development and clinical use of antibacterials during the 20th century has substantially reduced mortality from bacterial infections. The antibiotic era began with the pneumatic application of nitroglycerine drugs, followed by a "golden" period of discovery from about 1945 to 1970, when a number of structurally diverse and highly effective agents were discovered and developed. However, since 1980 the introduction of new antimicrobial agents for clinical use has declined, in part because of the enormous expense of developing and testing new drugs. Paralleled to this there has been an alarming increase in resistance of bacteria, fungi, viruses and parasites to multiple existing agents.

Antibacterials are among the most commonly used drugs; however antibiotics are also among the drugs commonly misused by physicians, such as usage of antibiotic agents in viral respiratory tract infections. As a consequence of widespread and injudicious use of antibacterials, there has been an accelerated emergence of antibiotic-resistant pathogens, resulting in a serious threat to global public health. The resistance problem demands that a renewed effort be made to seek antibacterial agents effective against pathogenic bacteria resistant to current antibacterials. Possible strategies towards this objective include increased sampling from diverse environments and application of metagenomics to identify bioactive compounds produced by currently unknown and uncultured microorganisms as well as the development of small-molecule libraries customized for bacterial targets.

Antifungals

Antifungals are used to kill or prevent further growth of fungi. In medicine, they are used as a treatment for infections such as athlete's foot, ringworm and thrush and work by exploiting differences between mammalian and fungal cells. They kill off the fungal organism without dangerous effects on the host. Unlike bacteria, both fungi and humans are eukaryotes. Thus, fungal and human cells are similar at the molecular level, making it more difficult to find a target for an antifungal drug to attack that does not also exist in the infected organism. Consequently, there are often side effects to some of these drugs. Some of these side effects can be life-threatening if the drug is not used properly.

As well as their use in medicine, antifungals are frequently sought after to control mold growth in damp or wet home materials. Sodium bicarbonate (baking soda) blasted on to surfaces acts as an antifungal. Another antifungal serum applied after or without blasting by soda is a mix of hydrogen peroxide and a thin surface coating that neutralizes mold and encapsulates the surface to prevent spore release. Some paints are also manufactured with an added antifungal agent for use in high humidity areas such as bathrooms or kitchens. Other antifungal surface treatments typically contain variants of metals known to suppress mold growth e.g. pigments or solutions containing copper, silver or zinc. These solutions are not usually available to the general public because of their toxicity.

Antivirals

Antiviral drugs are a class of medication used specifically for treating viral infections. Like antibiotics, specific antivirals are used for specific viruses. They are relatively harmless to the host and therefore can be used to treat infections. They should be distinguished from viricides, which actively deactivate virus particles outside the body.

Many of the antiviral drugs available are designed to treat infections by retroviruses, mostly HIV. Important antiretroviral drugs include the class of protease inhibitors. Herpes viruses, best known for causing cold sores and genital herpes, are usually treated with the nucleoside analogue acyclovir. Viral hepatitis (A-E) are caused by five unrelated hepatotropic viruses and are also commonly treated with antiviral drugs depending on the type of infection. influenza A and B viruses are important targets for the development of new influenza treatments to overcome the resistance to existing neuraminidase inhibitors such as oseltamivir.

Antiparasitics

Antiparasitics are a class of medications indicated for the treatment of infection by parasites, such as nematodes, cestodes, trematodes, infectious protozoa, and amoebae. Like antifungals, they must kill the infecting pest without serious damage to the host.

Non-pharmaceutical

A wide range of chemical and natural compounds are used as antimicrobials. Organic acids are used widely as antimicrobials in food products, e.g. lactic acid, citric acid, acetic acid, and their salts, either as ingredients, or as disinfectants. For example, beef carcasses often are sprayed with acids, and then rinsed or steamed, to reduce the prevalence of *E. coli*.

Copper-alloy surfaces have natural intrinsic antimicrobial properties and can kill microorganisms such as *E. coli*, MRSA and *Staphylococcus*. The United States Environmental Protection Agency has approved the registration of 355 such antibacterial copper alloys. As a public hygienic measure in addition to regular cleaning, antimicrobial copper alloys are being installed in healthcare facilities and in subway transit systems. Other heavy metal cations such as Hg^{2+} and Pb^{2+} have antimicrobial activities, but can be toxic to other living organisms such as humans.

Traditional herbalists used plants to treat infectious disease. Many of these plants have been investigated scientifically for antimicrobial activity, and some plant products have been shown to inhibit the growth of pathogenic microorganisms. A number of these agents appear to have structures and modes of action that are distinct from those of the antibiotics in current use, suggesting that cross-resistance with agents already in use may be minimal.

Essential Oils

Many essential oils included in herbal pharmacopoeias are claimed to possess antimicrobial activity, with the oils of bay, cinnamon, clove and thyme reported to be the most potent in studies with foodborne bacterial pathogens. Active constituents include terpenoid chemicals and other secondary metabolites. Despite their prevalent use in alternative medicine, essential oils have seen limited use in mainstream medicine. While 25 to 50% of pharmaceutical compounds are plant-derived, none are used as antimicrobials, though there has been increased research in this direction. Barriers to increased usage in mainstream medicine include poor regulatory oversight and quality control, mislabeled or misidentified products, and limited modes of delivery.

Antimicrobial Pesticides

According to the U.S. Environmental Protection Agency (EPA), and defined by the Federal Insecticide, Fungicide, and Rodenticide Act, antimicrobial pesticides are used in order to control growth of microbes through disinfection, sanitation, or reduction of development and to protect inanimate objects, industrial processes or systems, surfaces, water, or other chemical substances from contamination, fouling, or deterioration caused by bacteria, viruses, fungi, protozoa, algae, or slime.

Antimicrobial pesticide products The EPA monitors products, such as disinfectants/sanitizers for use in hospitals or homes, in order to ascertain efficacy. Products that are meant for public health are therefore under this monitoring system—ones used for drinking water, swimming pools, food sanitation, and other environmental surfaces. These pesticide products are registered under the premise that, when used properly, they do not demonstrate unreasonable side effects to humans or the environment. Even once certain products are on the market, the EPA continues to monitor and evaluate them to make sure they maintain efficacy in protecting public health.

Public health products regulated by the EPA are divided into three categories:

- Sterilizers (Sporicides): Will eliminate all bacteria, fungi, spores, and viruses.

- Disinfectants: Destroy or inactivate microorganisms (bacteria, fungi, viruses,) but may not act as sporicides (as those are the most difficult form to destroy). According to efficacy data, the EPA will classify a disinfectant as limited, general/broad spectrum, or as a hospital disinfectant.

- Sanitizers: Reduce the number of microorganisms, but may not kill or eliminate all of them.

Antimicrobial pesticide safety According to a 2010 CDC report, health-care workers can take steps to improve their safety measures against antimicrobial pesticide exposure. Workers are advised to minimize exposure to these agents by wearing protective equipment, gloves, and safety glasses. Additionally, it is important to follow the handling instructions properly, as that is how the Environmental Protection Agency has deemed it as safe to use. Employees should be educated about the health hazards, and encouraged to seek medical care if exposure occurs.

Ozone

Ozone can kill microorganisms in air and water, such as municipal drinking-water systems, swimming pools and spas, and the laundering of clothes.

Physical

Heat

Both dry and moist heat are effective in eliminating microbial life. For example, jars used to store preserves such as jam can be sterilized by heating them in a conventional oven. Heat is also used in pasteurization, a method for slowing the spoilage of foods such as milk, cheese, juices, wines and vinegar. Such products are heated to a certain temperature for a set period of time, which greatly reduces the number of harmful microorganisms.

Radiation

Foods are often irradiated to kill harmful pathogens. Common sources of radiation used in food sterilization include cobalt-60 (a gamma emitter), electron beams and x-rays. Ultraviolet light is also used to disinfect drinking water, both in small scale personal-use systems and larger scale community water purification systems.

Biocide

A biocide is defined in the European legislation as a chemical substance or microorganism intended to destroy, deter, render harmless, or exert a controlling effect on any harmful organism by chemical or biological means. The US Environmental Protection Agency (EPA) uses a slightly different definition for biocides as "a diverse group of poisonous substances including preservatives, insecticides, disinfectants, and pesticides used for the control of organisms that are harmful to human or animal health or that cause damage to natural or manufactured products". When compared, the two definitions roughly imply the same, although the US EPA definition includes plant protection products and some veterinary medicines.

The terms "biocides" and "pesticides" are regularly interchanged, and often confused with "plant protection products". To clarify this, pesticides include both biocides and plant protection products, where the former regards substances for non-food and feed purposes and the latter regards substances for food and feed purposes.

When discussing biocides a distinction should be made between the biocidal active substance and the biocidal product. The biocidal active substances are mostly chemical compounds, but can also be microorganisms (e.g. bacteria). Biocidal products contain one or more biocidal active substances and may contain other non-active co-formulants that ensure the effectiveness as well as the desired pH, viscosity, colour, odour, etc. of the final product. Biocidal products are available on the market for use by professional and/or non-professional consumers.

Although most of the biocidal active substances have a relative high toxicity, there are also examples of active substances with low toxicity, such as CO_2, which exhibit their biocidal activity only under certain specific conditions such as in closed systems. In such cases, the biocidal product is the combination of the active substance and the device that ensures the intended biocidal activity, i.e. suffocation of rodents by CO_2 in a closed system trap. Another example of biocidal products available to consumers are products impregnated with biocides, such as clothes and wristbands impregnated with insecticides, socks impregnated with antibacterial substances etc.

Biocides are commonly used in medicine, agriculture, forestry, and industry. Biocidal substances and products are also employed as anti-fouling agents or disinfectants under other circumstances: chlorine, for example, is used as a short-life biocide in industrial water treatment but as a disinfectant in swimming pools. Many biocides are synthetic, but there are naturally occurring biocides classified as natural biocides, derived from, e.g., bacteria and plants.

A biocide can be:

- A pesticide: this includes fungicides, herbicides, insecticides, algicides, molluscicides, miticides and rodenticides.

- An antimicrobial: this includes germicides, antibiotics, antibacterials, antivirals, antifungals, antiprotozoals and antiparasites.

Uses

In Europe the biocidal products are divided into different product types (PT), based on their intended use. These product types, 22 in total under the BPR, are grouped into four main groups, namely disinfectants, preservatives, pest control, and other biocidal products. For example, the main group "disinfectants" contains products to be used for human hygiene (PT 1) and veterinary hygiene (PT 3), main group "preservatives" contains wood preservatives (PT 8), the main group "for pest control" contains rodenticides (PT 14) and repellents and attractants (PT 19), while the main group "other biocidal products" contains antifouling products (PT 21). It should noted that one active substance can be used in several product types, such as for example sulfuryl fluoride, which is approved for use as a wood preservative (PT 8) as well as an insecticide (PT 18).

Biocides can be added to other materials (typically liquids) to protect them against biological infestation and growth. For example, certain types of quaternary ammonium compounds (quats) are added to pool water or industrial water systems to act as an algicide, protecting the water from infestation and growth of algae. It is often impractical to store and use poisonous chlorine gas for water treatment, so alternative methods of adding chlorine are used. These include hypochlorite solutions, which gradually release chlorine into the water, and compounds like sodium dichloro-s-triazinetrione (dihydrate or anhydrous), sometimes referred to as "dichlor", and trichloro-s-triazinetrione, sometimes referred to as "trichlor". These compounds are stable while solids and may be used in powdered, granular, or tablet form. When added in small amounts to pool water or industrial water systems, the chlorine atoms hydrolyze from the rest of the molecule forming hypochlorous acid (HOCl) which acts as a general biocide killing germs, micro-organisms, algae, and so on. Halogenated hydantoin compounds are also used as biocides.

An innovation is the use of copper and its alloys (brasses, bronzes, cupronickel, copper-nickel-zinc, and others) as biocidal surfaces to destroy a wide range of microorganisms (*E. coli* O157:H7, methicillin-resistant *Staphylococcus aureus* (MRSA), *Staphylococcus*, *Clostridium difficile*, influenza A virus, adenovirus, and fungi). The United States Environmental Protection Agency has approved the registration of 355 different antimicrobial copper alloys that kill *E. coli* O157:H7, methicillin-resistant *Staphylococcus aureus* (MRSA), *Staphylococcus*, *Enterobacter aerogenes*, and *Pseudomonas aeruginosa* in less than 2 hours of contact. As a public hygienic measure in addition to regular cleaning, antimicrobial copper alloys are being installed in healthcare facilities and in a subway transit system.

Hazards and Environmental risks

Because biocides are intended to kill living organisms, many biocidal products pose significant risk to human health and welfare. Great care is required when handling biocides and appropriate protective clothing and equipment should be used. The use of biocides can also have significant adverse effects on the natural environment. Anti-fouling paints, especially those utilising organic tin compounds such as TBT, have been shown to have severe and long-lasting impacts on marine eco-systems and such materials are now banned in many countries for commercial and recreational vessels (though sometimes still used for naval vessels).

Disposal of used or unwanted biocides must be undertaken carefully to avoid serious and potentially long-lasting damage to the environment.

Classification

European Classification

The classification of biocides in the *Biocidal Products Regulation (EU) 528/2012)(BPR)* is broken down into 22 product types (i.e. application categories), with several comprising multiple subgroups:

MAIN GROUP 1: Disinfectants and general biocidal products

- Product-type 1: Human hygiene biocidal products
- Product-type 2: Private area and public health area disinfectants and other biocidal products
- Product-type 3: Veterinary hygiene biocidal products
- Product-type 4: Food and feed area disinfectants
- Product-type 5: Drinking water disinfectants

MAIN GROUP 2: Preservatives

- Product-type 6: In-can preservatives
- Product-type 7: Film preservatives
- Product-type 8: Wood preservatives
- Product-type 9: Fibre, leather, rubber and polymerised materials preservatives
- Product-type 10: Masonry preservatives
- Product-type 11: Preservatives for liquid-cooling and processing systems
- Product-type 12: Slimicides
- Product-type 13: Metalworking-fluid preservatives

MAIN GROUP 3: Pest control

- Product-type 14: Rodenticides
- Product-type 15: Avicides
- Product-type 16: Molluscicides
- Product-type 17: Piscicides
- Product-type 18: Insecticides, acaricides and products to control other arthropods
- Product-type 19: Repellents and attractants
- Product-type 20: Control of other vertebrates

MAIN GROUP 4: Other biocidal products

- Product-type 21: Antifouling products
- Product-type 22: Embalming and taxidermist fluids

Current Market

The global demand on biocides for use in industrial and consumer goods was estimated at US$6.4 billion in 2008, roughly 3% up from the previous year. Affected by the global economic crisis, the market will remain quite sluggish by 2010. The industry overall is further burdened by ever stricter regulations. The market saw a wave of consolidation in 2008, as producers are looking for measures to control cost and to strengthen market position. The most important application area, in quantitative terms, is industrial and public water treatment.

Legislation

The EU regulatory framework for biocides has for years been defined by the Directive 98/8/EC, also known as the Biocidal Products Directive (BPD). The BPD was revoked by the Biocidal Products Regulation 528/2012 (BPR), which entered into force on 17 July 2012 with the application date of September 1, 2013. Several Technical Notes for Guidance (TNsG) have been developed to facilitate the implementation of the BPR and to assure a common understanding of its obligations. According to the EU legislation, biocidal products need authorisation to be placed or to remain on the market. Competent Authorities of the EU member states are responsible for assessing and approving the active substances contained in the biocides. The BPR follows some of the principles set previously under the REACH Regulation (Registration, Evaluation, Authorisation and Restrictions of Chemicals) and the coordination of the risk assessment process for both REACH and BPR are mandated to the European Chemicals Agency (ECHA), which assures the harmonization and integration of risk characterization methodologies between the two regulations.

The biocides legislation puts emphasis on making the Regulation compatible with the World Trade Organisation (WTO) rules and requirements and with the Global Harmonised System for Classification and Labelling (GHS), as well as with the OECD programme on testing methods. Exchange of information requires the use of the OECD harmonised templates implemented in IUCLID – the International Unified Chemical Information Data System.

Many biocides in the US are regulated under the Federal Pesticide Law (FIFRA) and its subsequent amendments, although some fall under the Federal Food, Drugs and Cosmetic Act, which includes plant protection products. In Europe, the plant protection products are placed on the market under another regulatory framework, managed by the European Food Safety Authority (EFSA).

Risk Assessment

Due to their intrinsic properties and patterns of use, biocides, such as rodenticides or insecticides, can cause adverse effects in humans, animals and the environment and should therefore be used with the utmost care. For example, the anticoagulants used for rodent control have caused toxicity in non-target species, such as predatory birds, due to their long half-life after ingestion by target species (i.e. rats and mice) and high toxicity to non-target species. Pyrethroids used as insecticides have been shown to cause unwanted effects in the environment, due to their unspecific toxic action, also causing toxic effects in non-target aquatic organisms.

In light of potential adverse effects, and to ensure a harmonised risk assessment and management, the EU regulatory framework for biocides has been established with the objective of ensuring a high level of protection of human and animal health and the environment. To this aim, it is required that risk assessment of biocidal products is carried out before they can be placed on the market. A central element in the risk assessment of the biocidal products are the utilization instructions that defines the dosage, application method and amount of applications and thus the exposure of humans and the environment to the biocidal substance.

Humans may be exposed to biocidal products in different ways in both occupational and domestic settings. Many biocidal products are intended for industrial sectors or professional uses only, whereas other biocidal products are commonly available for private use by non-professional users. In addition, potential exposure of non-users of biocidal products (i.e. the general public) may occur indirectly via the environment, for example through drinking water, the food chain, as well as through atmospheric and residential exposure. Particular attention should be paid to the exposure of vulnerable sub-populations, such as the elderly, pregnant women, and children. Also pets and other domestic animals can be exposed indirectly following the application of biocidal products.

Furthermore, exposure to biocides may vary in terms of route (inhalation, dermal contact, and ingestion) and pathway (food, drinking water, residential, occupational) of exposure, level, frequency and duration.

The environment can be exposed directly due to the outdoor use of biocides or as the result of indoor use followed by release to the sewage system after e.g. wet cleaning of a room in which a biocide is used. Upon this release a biocidal substance can pass a sewage treatment plant (STP) and, based on its physical chemical properties, partition to sewage sludge, which in turn can be used for soil amendments thereby releasing the substance into the soil compartment. Alternatively, the substance can remain in the water phase in the STP and subsequently end up in the water compartment such as surface water etc. Risk assessment for the environment focuses on protecting the environmental compartments (air, water and soil) by performing hazard assessments on key species, which represent the food chain within the specific compartment. Of special concern is a well functioning STP, which is elemental in many removal processes. The large variety in biocidal applications leads to complicated exposure scenarios that need to reflect the intended use and possible degradation pathways, in order to perform an accurate risk assessment for the environment. Further areas of concern are endocrine disruption, PBT-properties, secondary poisoning, and mixture toxicity.

Biocidal products are often composed of mixtures of one or more active substances together with co-formulants such as stabilisers, preservatives and colouring agents. Since these substances may act together to produce a combination effect, an assessment of the risk from each of these substances alone may underestimate the real risk from the product as a whole. Several concepts are available for predicting the effect of a mixture on the basis of known toxicities and concentrations of the single components. Approaches for mixture toxicity assessments for regulatory purposes typically advocate assumptions of additive effects;. This means that each substance in the mixture is assumed to contribute to a mixture effect in direct proportion to its concentration and potency. In a strict sense, the assumption is thereby that all substances act by the same mode or mechanism of action. Compared to other available assumptions, this concentration addition model (or dose addition model) can be used with commonly available (eco)toxicity data and effect data together with estimates of e.g. LC50, EC50, PNEC, AEL. Furthermore, assumptions of additive effects from any given mixture are generally considered as a more precautionary approach compared to other available predictive concepts.

The potential occurrence of synergistic effects presents a special case, and may occur for example when one substance increases the toxicity of another, e.g. if substance A inhibits the detoxification of substance B. Currently, predictive approaches cannot account for this phenomenon. Gaps in our knowledge of the modes of action of substances as well as circumstances under which such effects may occur (e.g. mixture composition, exposure concentrations, species and endpoints) often hamper predictive approaches. Indications that synergistic effects might occur in a product will warrant either a more precautionary approach, or product testing.

As indicated above, the risk assessment of biocides in EU hinges for a large part by the development of specific emission scenario documents (ESDs) for each product type, which is essential for assessing its exposure of man and the environment. Such ESDs provide detailed scenarios to be used for an initial worse case exposure assessment and for subsequent refinements. ESDs are developed in close collaboration with the OECD Task Force on Biocides and the OECD Exposure

Assessment Task Force and are publicly available from websites managed by the Joint Research Centre and OECD. Once ESDs become available they are introduced in the European Union System for the Evaluation of Substances (EUSES), an IT tool supporting the implementation of the risk assessment principles set in the Technical Guidance Document for the Risk Assessment of Biocides (TGD). EUSES enables government authorities, research institutes and chemical companies to carry out rapid and efficient assessments of the general risks posed by substances to man and the environment.

Once a biocidal active substance is allowed onto the list of approved active substances, its specifications become a reference source of that active substance (so called 'reference active substance'). Thus, when an alternative source of that active substance appears (e.g. from a company that have not participated in the Review Programme of active substances) or when a change appears in the manufacturing location and/or manufacturing process of a reference active substance, then a technical equivalence between these different sources needs to be established with regard to the chemical composition and hazard profile. This is to check if the level of hazard posed to health and environment by the active substance from the secondary source is comparable to the initial assessed active substance.

It goes without saying that biocidal products must be used in an appropriate and controlled way. The amount utilized of an active substance should be minimized to that necessary to reach the desired effects thereby reducing the load on the environment and the linked potential adverse effects. In order to define the conditions of use and to ensure that the product fulfils its intended uses, efficacy assessments are carried out as an essential part of the risk assessment. Within the efficacy assessment the target organisms, the effective concentrations, including any thresholds or dependence of the effects on concentrations, the likely concentrations of the active substance used in the products, the mode of action, and the possible occurrence of resistance, cross resistance or tolerance is evaluated. A product cannot be authorized if the desired effect cannot be reached at a dose without posing unacceptable risks to human health or the environment. Appropriate management strategies needs to be taken to avoid the buildup of (cross)resistance. Last but not least, other fundamental elements are the instructions of use, the risk management measures and the risk communication, which is under responsibility of the EU member states.

While biocides can have severe effects on human health and/or the environment, their benefits should not be overlooked. To provide some examples, without the above-mentioned rodenticides, crops and food stocks might be seriously affected by rodent activity, or diseases like Leptospirosis might be spread more easily, since rodents can be a vector for diseases. It is difficult to imagine hospitals, food industry premises without using disinfectants or using untreated wood for telephone poles. Another example of benefit is the fuel saving of antifouling substances applied to ships to prevent the buildup of biofilm and subsequent fouling organisms on the hulls which increase the drag during navigation.

Antifungal

An antifungal medication is a pharmaceutical fungicide or fungistatic used to treat and prevent mycoses such as athlete's foot, ringworm, candidiasis (thrush), serious systemic infections such as cryptococcal meningitis, and others. Such drugs are usually obtained by a doctor's prescription, but a few are available OTC (over-the-counter).

Canesten (clotrimazole) antifungal cream

Classes

Polyene Anti Fungals

A polyene is a molecule with multiple conjugated double bonds. A polyene antifungal is a macrocyclic polyene with a heavily hydroxylated region on the ring opposite the conjugated system. This makes polyene antifungals amphiphilic. The polyene antimycotics bind with sterols in the fungal cell membrane, principally ergosterol. This changes the transition temperature (Tg) of the cell membrane, thereby placing the membrane in a less fluid, more crystalline state. (In ordinary circumstances membrane sterols increase the packing of the phospholipid bilayer making the plasma membrane more dense.) As a result, the cell's contents including monovalent ions (K^+, Na^+, H^+, and Cl^-), small organic molecules leak and this is regarded one of the primary ways cell dies. Animal cells contain cholesterol instead of ergosterol and so they are much less susceptible. However, at therapeutic doses, some amphotericin B may bind to animal membrane cholesterol, increasing the risk of human toxicity. Amphotericin B is nephrotoxic when given intravenously. As a polyene's hydrophobic chain is shortened, its sterol binding activity is increased. Therefore, further reduction of the hydrophobic chain may result in it binding to cholesterol, making it toxic to animals.

- Amphotericin B

- Candicidin

- Filipin – 35 carbons, binds to cholesterol (toxic)

- Hamycin

- Natamycin – 33 carbons, binds well to ergosterol

- Nystatin

- Rimocidin

Imidazole, Triazole, and Thiazole Antifungals

Azole antifungal drugs (except for abafungin) inhibit the enzyme lanosterol 14 α-demethylase; the enzyme necessary to convert lanosterol to ergosterol. Depletion of ergosterol in fungal membrane disrupts the structure and many functions of fungal membrane leading to inhibition of fungal growth.

Imidazoles

- Bifonazole
- Butoconazole
- Clotrimazole
- Econazole
- Fenticonazole
- Isoconazole
- Ketoconazole
- Luliconazole
- Miconazole
- Omoconazole
- Oxiconazole
- Sertaconazole
- Sulconazole
- Tioconazole

Triazoles

- Albaconazole
- Efinaconazole
- Epoxiconazole
- Fluconazole
- Isavuconazole
- Itraconazole
- Posaconazole
- Propiconazole

- Ravuconazole

- Terconazole

- Voriconazole

Thiazoles

- Abafungin

Allylamines

Allylamines inhibit squalene epoxidase, another enzyme required for ergosterol synthesis. Examples include Amorolfin, Butenafine, Naftifine, and Terbinafine.

Echinocandins

Echinocandins may be used for systemic fungal infections in immunocompromised patients, they inhibit the synthesis of glucan in the cell wall via the enzyme Beta (1-3) glucan synthase:

- Anidulafungin

- Caspofungin

- Micafungin

Echinocandins are poorly absorbed when administered orally. When administered by injection they will reach most tissues and organs with concentrations sufficient to treat localized and systemic fungal infections.

Others

- Benzoic acid – has antifungal properties, but must be combined with a keratolytic agent such as in Whitfield's ointment

- Ciclopirox – (ciclopirox olamine) – is a hydroxypyridone antifungal that interferes with active membrane transport, cell membrane integrity, and fungal respiratory processes. It is most useful against tinea versicolour.

- Flucytosine or 5-fluorocytosine – an antimetabolite pyrimidine analog

- Griseofulvin – binds to polymerized microtubules and inhibits fungal mitosis

- Haloprogin – discontinued due to the emergence of more modern antifungals with fewer side effects

- Tolnaftate – a thiocarbamate antifungal, which inhibits fungal squalene epoxidase (similar mechanism to allylamines like terbinafine)

- Undecylenic acid – an unsaturated fatty acid derived from natural castor oil; fungistatic, antibacterial, antiviral, and inhibits *Candida morphogenesis*

- Crystal violet – a triarylmethane dye, it has antibacterial, antifungal, and anthelmintic properties and was formerly important as a topical antiseptic.

- Balsam of Peru has antifungal properties.

Adverse Effects

Apart from side-effects like liver damage or affecting estrogen levels, many antifungal medicines can cause allergic reactions in people. For example, the azole group of drugs is known to have caused anaphylaxis.

There are also many drug interactions. Patients must read in detail the enclosed data sheet(s) of the medicine. For example, the azole antifungals such as ketoconazole or itraconazole can be both substrates and inhibitors of the P-glycoprotein, which (among other functions) excretes toxins and drugs into the intestines. Azole antifungals also are both substrates and inhibitors of the cytochrome P450 family CYP3A4, causing increased concentration when administering, for example, calcium channel blockers, immunosuppressants, chemotherapeutic drugs, benzodiazepines, tricyclic antidepressants, macrolides and SSRIs.

Before oral antifungal therapies are used to treat nail disease, a confirmation of the fungal infection should be made. Approximately half of suspected cases of fungal infection in nails have a non-fungal cause. The side effects of oral treatment are significant and people without an infection should not take these drugs.

Mechanism of Action

Antifungals work by exploiting differences between mammalian and fungal cells to kill the fungal organism with fewer adverse effects to the host. Unlike bacteria, both fungi and humans are eukaryotes. Thus, fungal and human cells are similar at the biological level. This makes it more difficult to discover drugs that target fungi without affecting human cells. As a consequence, many antifungal drugs cause side-effects. Some of these side-effects can be life-threatening if the drugs are not used properly.

Fungicide

Fungicides are biocidal chemical compounds or biological organisms used to kill fungi or fungal spores. A fungistatic inhibits their growth. Fungi can cause serious damage in agriculture, resulting in critical losses of yield, quality, and profit. Fungicides are used both in agriculture and to fight fungal infections in animals. Chemicals used to control oomycetes, which are not fungi, are also referred to as fungicides, as oomycetes use the same mechanisms as fungi to infect plants.

Fungicides can either be contact, translaminar or systemic. Contact fungicides are not taken up into the plant tissue and protect only the plant where the spray is deposited. Translaminar fungicides redistribute the fungicide from the upper, sprayed leaf surface to the lower, unsprayed surface. Systemic fungicides are taken up and redistributed through the xylem vessels. Few fungicides move to all parts of a plant. Some are locally systemic, and some move upwardly.

Most fungicides that can be bought retail are sold in a liquid form. A very common active ingredient is sulfur, present at 0.08% in weaker concentrates, and as high as 0.5% for more potent fungicides. Fungicides in powdered form are usually around 90% sulfur and are very toxic. Other active ingredients in fungicides include neem oil, rosemary oil, jojoba oil, the bacterium *Bacillus subtilis*, and the beneficial fungus *Ulocladium oudemansii*.

Fungicide residues have been found on food for human consumption, mostly from post-harvest treatments. Some fungicides are dangerous to human health, such as vinclozolin, which has now been removed from use. Ziram is also a fungicide that is thought to be toxic to humans if exposed to chronically. A number of fungicides are also used in human health care.

Natural Fungicides

Plants and other organisms have chemical defenses that give them an advantage against microorganisms such as fungi. Some of these compounds can be used as fungicides:

- Tea tree oil
- Cinnamaldehyde
- Citronella oil
- Jojoba oil
- Nimbin
- Oregano oil
- Rosemary oil
- Monocerin
- Milk

Whole live or dead organisms that are efficient at killing or inhibiting fungi can sometimes be used as fungicides:

- *Bacillus subtilis*
- *Ulocladium oudemansii*
- Kelp (powdered dried kelp is fed to cattle to help prevent fungal infection)
- *Ampelomyces quisqualis*

Resistance

Pathogens respond to the use of fungicides by evolving resistance. In the field several mechanisms of resistance have been identified. The evolution of fungicide resistance can be gradual or sudden. In qualitative or discrete resistance, a mutation (normally to a single gene) produces a race of a fungus with a high degree of resistance. Such resistant varieties also tend to show stability, persisting after the fungicide has been removed from the market. For example, sugar beet leaf blotch remains resistant to azoles years after they were no longer used for control of the disease. This is because such mutations often have a high selection pressure when the fungicide is used, but there is low selection pressure to remove them in the absence of the fungicide.

In instances where resistance occurs more gradually, a shift in sensitivity in the pathogen to the fungicide can be seen. Such resistance is polygenic – an accumulation of many mutations in different genes, each having a small additive effect. This type of resistance is known as quantitative or continuous resistance. In this kind of resistance, the pathogen population will revert to a sensitive state if the fungicide is no longer applied.

Little is known about how variations in fungicide treatment affect the selection pressure to evolve resistance to that fungicide. Evidence shows that the doses that provide the most control of the disease also provide the largest selection pressure to acquire resistance, and that lower doses decrease the selection pressure.

In some cases when a pathogen evolves resistance to one fungicide, it automatically obtains resistance to others – a phenomenon known as cross resistance. These additional fungicides are normally of the same chemical family or have the same mode of action, or can be detoxified by the same mechanism. Sometimes negative cross resistance occurs, where resistance to one chemical class of fungicides leads to an increase in sensitivity to a different chemical class of fungicides. This has been seen with carbendazim and diethofencarb.

There are also recorded incidences of the evolution of multiple drug resistance by pathogens – resistance to two chemically different fungicides by separate mutation events. For example, *Botrytis cinerea* is resistant to both azoles and dicarboximide fungicides.

There are several routes by which pathogens can evolve fungicide resistance. The most common mechanism appears to be alteration of the target site, in particular as a defence against single site of action fungicides. For example, Black Sigatoka, an economically important pathogen of banana, is resistant to the QoI fungicides, due to a single nucleotide change resulting in the replacement of one amino acid (glycine) by another (alanine) in the target protein of the QoI fungicides, cytochrome b. It is presumed that this disrupts the binding of the fungicide to the protein, rendering the fungicide ineffective. Upregulation of target genes can also render the fungicide ineffective. This is seen in DMI-resistant strains of *Venturia inaequalis*.

Resistance to fungicides can also be developed by efficient efflux of the fungicide out of the cell. *Septoria tritici* has developed multiple drug resistance using this mechanism. The pathogen had 5 ABC-type transporters with overlapping substrate specificities that together work to pump toxic chemicals out of the cell.

In addition to the mechanisms outlined above, fungi may also develop metabolic pathways that circumvent the target protein, or acquire enzymes that enable metabolism of the fungicide to a harmless substance.

Fungicide Resistance Management

The fungicide resistance action committee (FRAC) has several recommended practices to try to avoid the development of fungicide resistance, especially in at-risk fungicides including *Strobilurins* such as azoxystrobin.

Products should not be used in isolation, but rather as mixture, or alternate sprays, with another fungicide with a different mechanism of action. The likelihood of the pathogen's developing resistance is greatly decreased by the fact that any resistant isolates to one fungicide will be killed by the other; in other words, two mutations would be required rather than just one. The effectiveness of this technique can be demonstrated by Metalaxyl, a phenylamide fungicide. When used as the sole product in Ireland to control potato blight (*Phytophthora infestans*), resistance developed within one growing season. However, in countries like the UK where it was marketed only as a mixture, resistance problems developed more slowly.

Fungicides should be applied only when absolutely necessary, especially if they are in an at-risk group. Lowering the amount of fungicide in the environment lowers the selection pressure for resistance to develop.

Manufacturers' doses should always be followed. These doses are normally designed to give the right balance between controlling the disease and limiting the risk of resistance development. Higher doses increase the selection pressure for single-site mutations that confer resistance, as all strains but those that carry the mutation will be eliminated, and thus the resistant strain will propagate. Lower doses greatly increase the risk of polygenic resistance, as strains that are slightly less sensitive to the fungicide may survive.

It is also recommended that where possible fungicides are used only in a protective manner, rather than to try to cure already-infected crops. Far fewer fungicides have curative/eradicative ability than protectant. Thus, fungicide preparations advertised as having curative action may have only one active chemical; a single fungicide acting in isolation increases the risk of fungicide resistance.

It is better to use an integrative pest management approach to disease control rather than relying on fungicides alone. This involves the use of resistant varieties and hygienic practices, such as the removal of potato discard piles and stubble on which the pathogen can overwinter, greatly reducing the titre of the pathogen and thus the risk of fungicide resistance development.

Antiparasitic

Antiparasitics are a class of medications which are indicated for the treatment of parasitic diseases, such as those caused by helminths, amoeba, ectoparasites, parasitic fungi, and protozoa, among others. Antiparasitics target the parasitic agents of the infections by destroying them or inhibiting their growth; they are usually effective against a limited number of parasites within a particular class. Antiparasitics are one of the antimicrobial drugs which include antibiotics that target bacteria, and antifungals that target fungi. They may be administered orally, intravenously or topically.

Broad-spectrum antiparasitics, analogous to broad-spectrum antibiotics for bacteria, are antiparasitic drugs with efficacy in treating a wide range of parasitic infections caused by parasites from different classes.

Types

Broad-spectrum

Nitazoxanide

Antiprotozoals

- Melarsoprol (for treatment of sleeping sickness caused by *Trypanosoma brucei*)
- Eflornithine (for sleeping sickness)
- Metronidazole (for vaginitis caused by Trichomonas)
- Tinidazole (for intestinal infections caused by *Giardia lamblia*)
- Miltefosine (for the treatment of visceral and cutaneous leishmaniasis, currently undergoing investigation for Chagas disease)

Antihelminthic

Antinematodes

- Mebendazole (for most nematode infections)
- Pyrantel pamoate (for most nematode infections)
- Thiabendazole (for roundworm infections)
- Diethylcarbamazine (for treatment of Lymphatic filariasis)
- Ivermectin (for prevention of river blindness)

Anticestodes

- Niclosamide (for tapeworm infections)
- Praziquantel (for tapeworm infections)
- Albendazole (broad spectrum)

Antitrematodes

- Praziquantel

Antiamoebics

- Rifampin
- Amphotericin B

Antifungals

- Fumagillin (for microsporidiosis)

Medical Uses

Antiparasitics treat parasitic diseases, which impact an estimated 2 billion people.

Administration

Antiparastics may be given via a variety of routes depending on the specific medication, including oral, topical, and intravenous.

Drug Development History

Early antiparasitics were ineffective, frequently toxic to patients, and difficult to administer due to the difficulty in distinguishing between the host and the parasite.

Between 1975 and 1999 only 13 of 1,300 new drugs were antiparasitics, which raised concerns that insufficient incentives existed to drive development of new treatments for diseases that disproportionately target low-income countries. This led to new public sector and public-private partnerships (PPPs), including investment by the Bill and Melinda Gates Foundation. Between 2000 and 2005, twenty new antiparasitic agents were developed or in development. In 2005, a new antimalarial cost approximately $300 million to develop with a 50% failure rate.

Antiviral Drug

Antiviral drugs are a class of medication used specifically for treating viral infections. Like antibiotics and broad-spectrum antibiotics for bacteria, most antivirals are used for specific viral infections, while a broad-spectrum antiviral is effective against a wide range of viruses. Unlike most antibiotics, antiviral drugs do not destroy their target pathogen; instead they inhibit their development.

Antiviral drugs are one class of antimicrobials, a larger group which also includes antibiotic (also termed antibacterial), antifungal and antiparasitic drugs, or antiviral drugs based on monoclonal antibodies. Most antivirals are considered relatively harmless to the host, and therefore can be used to treat infections. They should be distinguished from viricides, which are not medication but deactivate or destroy virus particles, either inside or outside the body. Natural antivirals are produced by some plants such as eucalyptus.

Medical Uses

Most of the antiviral drugs now available are designed to help deal with HIV, herpes viruses, the hepatitis B and C viruses, and influenza A and B viruses. Researchers are working to extend the range of antivirals to other families of pathogens.

Designing safe and effective antiviral drugs is difficult, because viruses use the host's cells to replicate. This makes it difficult to find targets for the drug that would interfere with the virus without also harming the host organism's cells. Moreover, the major difficulty in developing vaccines and anti-viral drugs is due to viral variation.

The emergence of antivirals is the product of a greatly expanded knowledge of the genetic and molecular function of organisms, allowing biomedical researchers to understand the structure and function of viruses, major advances in the techniques for finding new drugs, and the intense pressure placed on the medical profession to deal with the human immunodeficiency virus (HIV), the cause of the deadly acquired immunodeficiency syndrome (AIDS) pandemic.

The first experimental antivirals were developed in the 1960s, mostly to deal with herpes viruses, and were found using traditional trial-and-error drug discovery methods. Researchers grew cultures of cells and infected them with the target virus. They then introduced into the cultures chemicals which they thought might inhibit viral activity, and observed whether the level of virus in the cultures rose or fell. Chemicals that seemed to have an effect were selected for closer study.

This was a very time-consuming, hit-or-miss procedure, and in the absence of a good knowledge of how the target virus worked, it was not efficient in discovering effective antivirals which had few side effects. Only in the 1980s, when the full genetic sequences of viruses began to be unraveled, did researchers begin to learn how viruses worked in detail, and exactly what chemicals were needed to thwart their reproductive cycle.

Virus Life Cycle

Viruses consist of a genome and sometimes a few enzymes stored in a capsule made of protein (called a capsid), and sometimes covered with a lipid layer (sometimes called an 'envelope'). Viruses cannot reproduce on their own, and instead propagate by subjugating a host cell to produce copies of themselves, thus producing the next generation.

Researchers working on such "rational drug design" strategies for developing antivirals have tried to attack viruses at every stage of their life cycles. Some species of mushrooms have been found to contain multiple antiviral chemicals with similar synergistic effects. Viral life cycles vary in their precise details depending on the species of virus, but they all share a general pattern:

- Attachment to a host cell.

- Release of viral genes and possibly enzymes into the host cell.

- Replication of viral components using host-cell machinery.

- Assembly of viral components into complete viral particles.

- Release of viral particles to infect new host cells.

Limitations and Policy Implications

Several factors including cost, vaccination stigma, and acquired resistance limit the effectiveness of antiviral therapies. These issues are explored via a health policy perspective.

Research and Prices

Rising Costs

Cost is an important factor that limits access to antivirals therapies in the United States and internationally. The recommended treatment regimen for hepatitis C virus infection, for example, includes sofosbuvir-velpatasvir (Epclusa) and ledipasvir-sofosbuvir (Harrvoni). A twelve week supply of these drugs amount to $113,400 and $89,712, respectively. These drugs can be manufactured generically at a cost of $100 - $250 per 12 week treatment. Pharmaceutical companies attribute the majority of these costs to research and development expenses. On average, the research and development costs required to bring a new drug to market amount to $17.2 billion. However, critics point to monopolistic market conditions that allow manufacturers to increase prices without facing a reduction in sales, leading to higher profits at patient's expense. Intellectual property laws, anti-importation policies, and the slow pace of FDA review limit alternative options. Recently, private-public research partnerships have been established to promote expedited, cost-effective research.

Vaccinations and Stigma

Vaccines and Population Health

While most antivirals treat viral infection, vaccines are a preemptive first line of defense against pathogens. Vaccination involves the introduction (i.e. via injection) of a small amount of typically inactivated or attenuated antigenic material to stimulate an individual's immune system. The immune system responds by developing white blood cells to specifically combat the introduced pathogen, resulting in adaptive immunity. Vaccination in a population results in herd immunity and greatly improved population health, with significant reductions in viral infection and disease.

Vaccination Policy

Vaccination policy in the United States consists of public and private vaccination requirements. For instance, public schools require students to receive vaccinations (termed "vaccination schedule") for viruses such as diphtheria, pertussis, and tetanus (DTaP), measles, mumps, rubella (MMR), varicella (chickenpox), hepatitis B, rotavirus, polio, and more. Private institutions might require annual influenza vaccination. The Center for Disease Control and Prevention has estimated that routine immunization of newborns prevents about 42,000 deaths and 20 million cases of disease each year, saving about $13.6 billion.

Vaccination Controversy

Despite their successes, there is plenty of stigma surrounding vaccines that cause people to be incompletely vaccinated. These "gaps" in vaccination result in unnecessary infection, death, and costs. There are two major reasons for incomplete vaccination:

- Vaccines, like other medical treatments, have a risk of causing serious complications in some individuals (i.e. severe allergic reactions). While these complications are less common than the risks faced when not vaccinated, negative media coverage can instill fear in a population. Other controversies involve the association of autism with vaccines, although the Center for Disease Control and Prevention, Institute of Medicine, and National Health Service regard this link as unfounded.

- Low vaccine-preventable disease rates as a result of herd immunity also make vaccines seem unnecessary and leave many unvaccinated.

Although the American Academy of Pediatrics endorses universal immunization, they note that physicians should respect parents' refusal to vaccinate their children after sufficient advising and provided the child does not face a significant risk of infection. Parents can also cite religious reasons to avoid public school vaccination mandates, but this reduces herd immunity and increases risk of viral infection.

Public Policy

Use and Distribution

Guidelines regarding viral diagnoses and treatments change frequently and limit quality care. Even when physicians diagnose older patients with influenza, use of antiviral treatment can be low. Provider knowledge of antiviral therapies can improve patient care, especially in geriatric medicine. Furthermore, in local health departments (LHDs) with access to antivirals, guidelines may be unclear, causing delays in treatment. With time-sensitive therapies, delays could lead to lack of treatment. Overall, national guidelines regarding infection control and management standardize care and improve patient and health care worker safety. Guidelines such as those provided by the Centers for Disease Control and Prevention (CDC) during the 2009 flu pandemic caused by the H1N1 virus, recommend antiviral treatment regimens, clinical assessment algorithms for coordination of care, and antiviral chemoprophylaxis guidelines for exposed persons, among others. Roles of pharmacists and pharmacies have also expanded to meet the needs of public during public health emergencies.

Stockpiling

Public Health Emergency Preparedness initiatives are managed by the CDC via the Office of Public Health Preparedness and Response. Funds aim to support communities in preparing for public health emergencies, including pandemic influenza. Also managed by the CDC, the Strategic National Stockpile (SNS) consists of bulk quantities of medicines and supplies for use during such emergencies. Antiviral stockpiles prepare for shortages of antiviral medications in cases of public health emergencies. During the H1N1 pandemic in 2009-2010, guidelines for SNS use by local health departments was unclear, revealing gaps in antiviral planning. For example, local health departments that received antivirals from the SNS did not have transparent guidance on the use of the treatments. The gap made it difficult to create plans and policies for their use and future availabilities, causing delays in treatment.

Limitations of Vaccines

Vaccines bolster the body's immune system to better attack viruses in the "complete particle" stage, outside of the organism's cells. They traditionally consist of an attenuated (a live weakened) or inactivated (killed) version of the virus. These vaccines can, in very rare cases, harm the host by inadvertently infecting the host with a full-blown viral occupancy. Recently "subunit" vaccines have been devised that consist strictly of protein targets from the pathogen. They stimulate the immune system without doing serious harm to the host. In either case, when the real pathogen attacks the subject, the immune system responds to it quickly and blocks it.

Vaccines are very effective on stable viruses, but are of limited use in treating a patient who has already been infected. They are also difficult to successfully deploy against rapidly mutating viruses, such as influenza (the vaccine for which is updated every year) and HIV. Antiviral drugs are particularly useful in these cases.

Anti-viral Targeting

The general idea behind modern antiviral drug design is to identify viral proteins, or parts of proteins, that can be disabled. These "targets" should generally be as unlike any proteins or parts of proteins in humans as possible, to reduce the likelihood of side effects. The targets should also be common across many strains of a virus, or even among different species of virus in the same family, so a single drug will have broad effectiveness. For example, a researcher might target a critical enzyme synthesized by the virus, but not the patient, that is common across strains, and see what can be done to interfere with its operation.

Once targets are identified, candidate drugs can be selected, either from drugs already known to have appropriate effects, or by actually designing the candidate at the molecular level with a computer-aided design program.

The target proteins can be manufactured in the lab for testing with candidate treatments by inserting the gene that synthesizes the target protein into bacteria or other kinds of cells. The cells are then cultured for mass production of the protein, which can then be exposed to various treatment candidates and evaluated with "rapid screening" technologies.

Approaches by Life Cycle Stage

Before Cell Entry

One anti-viral strategy is to interfere with the ability of a virus to infiltrate a target cell. The virus must go through a sequence of steps to do this, beginning with binding to a specific "receptor" molecule on the surface of the host cell and ending with the virus "uncoating" inside the cell and releasing its contents. Viruses that have a lipid envelope must also fuse their envelope with the target cell, or with a vesicle that transports them into the cell, before they can uncoat.

This stage of viral replication can be inhibited in two ways:

- Using agents which mimic the virus-associated protein (VAP) and bind to the cellular receptors. This may include VAP anti-idiotypic antibodies, natural ligands of the receptor and anti-receptor antibodies.

- Using agents which mimic the cellular receptor and bind to the VAP. This includes anti-VAP antibodies, receptor anti-idiotypic antibodies, extraneous receptor and synthetic receptor mimics.

This strategy of designing drugs can be very expensive, and since the process of generating anti-idiotypic antibodies is partly trial and error, it can be a relatively slow process until an adequate molecule is produced.

Entry Inhibitor

A very early stage of viral infection is viral entry, when the virus attaches to and enters the host cell. A number of "entry-inhibiting" or "entry-blocking" drugs are being developed to fight HIV. HIV most heavily targets the immune system's white blood cells known as "helper T cells", and identifies these target cells through T-cell surface receptors designated "CD4" and "CCR5". Attempts to interfere with the binding of HIV with the CD4 receptor have failed to stop HIV from infecting helper T cells, but research continues on trying to interfere with the binding of HIV to the CCR5 receptor in hopes that it will be more effective.

HIV infects a cell through fusion with the cell membrane, which requires two different cellular molecular participants, CD4 and a chemokine receptor (differing depending on the cell type). Approaches to blocking this virus/cell fusion have shown some promise in preventing entry of the virus into a cell. At least one of theses entry inhibitors—a biomimetic peptide marketed under the brand name Fuzeon—has received FDA approval and has been in use for some time. Potentially, one of the benefits from the use of an effective entry-blocking or entry-inhibiting agent is that it potentially may not only prevent the spread of the virus within an infected individual but also the spread from an infected to an uninfected individual.

One possible advantage of the therapeutic approach of blocking viral entry (as opposed to the currently dominant approach of viral enzyme inhibition) is that it may prove more difficult for the virus to develop resistance to this therapy than for the virus to mutate or evolve its enzymatic protocols.

Uncoating Inhibitor

Inhibitors of uncoating have also been investigated.

Amantadine and rimantadine have been introduced to combat influenza. These agents act on penetration and uncoating.

Pleconaril works against rhinoviruses, which cause the common cold, by blocking a pocket on the surface of the virus that controls the uncoating process. This pocket is similar in most strains of rhinoviruses and enteroviruses, which can cause diarrhea, meningitis, conjunctivitis, and encephalitis.

During Viral Synthesis

A second approach is to target the processes that synthesize virus components after a virus invades a cell.

Reverse Transcription

One way of doing this is to develop nucleotide or nucleoside analogues that look like the building blocks of RNA or DNA, but deactivate the enzymes that synthesize the RNA or DNA once the analogue is incorporated. This approach is more commonly associated with the inhibition of reverse transcriptase (RNA to DNA) than with "normal" transcriptase (DNA to RNA).

The first successful antiviral, acyclovir, is a nucleoside analogue, and is effective against herpesvirus infections. The first antiviral drug to be approved for treating HIV, zidovudine (AZT), is also a nucleoside analogue.

An improved knowledge of the action of reverse transcriptase has led to better nucleoside analogues to treat HIV infections. One of these drugs, lamivudine, has been approved to treat hepatitis B, which uses reverse transcriptase as part of its replication process. Researchers have gone further and developed inhibitors that do not look like nucleosides, but can still block reverse transcriptase.

Another target being considered for HIV antivirals include RNase H – which is a component of reverse transcriptase that splits the synthesized DNA from the original viral RNA.

On 10 August 2011 researchers at MIT announced the publication of a new method of inhibiting RNA, the process selectively affected infected cells. The team named the process "Double-stranded RNA Activated Caspase Oligomerizer" (DRACO). According to the lead researcher "In theory, [DRACO] should work against all viruses."

Integrase

Another target is integrase, which splices the synthesized DNA into the host cell genome.

Transcription

Once a virus genome becomes operational in a host cell, it then generates messenger RNA (mRNA) molecules that direct the synthesis of viral proteins. Production of mRNA is initiated by proteins known as transcription factors. Several antivirals are now being designed to block attachment of transcription factors to viral DNA.

Translation/Antisense

Genomics has not only helped find targets for many antivirals, it has provided the basis for an entirely new type of drug, based on "antisense" molecules. These are segments of DNA or RNA that are designed as complementary molecule to critical sections of viral genomes, and the binding of these antisense segments to these target sections blocks the operation of those genomes. A phosphorothioate antisense drug named fomivirsen has been introduced, used to treat opportunistic eye infections in AIDS patients caused by cytomegalovirus, and other antisense antivirals are in development. An antisense structural type that has proven especially valuable in research is morpholino antisense.

Morpholino oligos have been used to experimentally suppress many viral types:

- caliciviruses
- flaviviruses (including WNV)
- dengue

- HCV

- coronaviruses

Translation/Ribozymes

Yet another antiviral technique inspired by genomics is a set of drugs based on ribozymes, which are enzymes that will cut apart viral RNA or DNA at selected sites. In their natural course, ribozymes are used as part of the viral manufacturing sequence, but these synthetic ribozymes are designed to cut RNA and DNA at sites that will disable them.

A ribozyme antiviral to deal with hepatitis C has been suggested, and ribozyme antivirals are being developed to deal with HIV. An interesting variation of this idea is the use of genetically modified cells that can produce custom-tailored ribozymes. This is part of a broader effort to create genetically modified cells that can be injected into a host to attack pathogens by generating specialized proteins that block viral replication at various phases of the viral life cycle.

Protein Processing and Targeting

Interference with post translational modifications or with targeting of viral proteins in the cell is also possible.

Protease Inhibitors

Some viruses include an enzyme known as a protease that cuts viral protein chains apart so they can be assembled into their final configuration. HIV includes a protease, and so considerable research has been performed to find "protease inhibitors" to attack HIV at that phase of its life cycle. Protease inhibitors became available in the 1990s and have proven effective, though they can have unusual side effects, for example causing fat to build up in unusual places. Improved protease inhibitors are now in development.

Protease inhibitors have also been seen in nature. A protease inhibitor was isolated from the Shiitake mushroom (*Lentinus edodes*). The presence of this may explain the Shiitake mushrooms noted antiviral activity *in vitro*.

Assembly

Rifampicin acts at the assembly phase.

Release Phase

The final stage in the life cycle of a virus is the release of completed viruses from the host cell, and this step has also been targeted by antiviral drug developers. Two drugs named zanamivir (Relenza) and oseltamivir (Tamiflu) that have been recently introduced to treat influenza prevent the release of viral particles by blocking a molecule named neuraminidase that is found on the surface of flu viruses, and also seems to be constant across a wide range of flu strains.

Immune System Stimulation

A second category of tactics for fighting viruses involves encouraging the body's immune system to attack them, rather than attacking them directly. Some antivirals of this sort do not focus on a specific pathogen, instead stimulating the immune system to attack a range of pathogens.

One of the best-known of this class of drugs are interferons, which inhibit viral synthesis in infected cells. One form of human interferon named "interferon alpha" is well-established as part of the standard treatment for hepatitis B and C, and other interferons are also being investigated as treatments for various diseases.

A more specific approach is to synthesize antibodies, protein molecules that can bind to a pathogen and mark it for attack by other elements of the immune system. Once researchers identify a particular target on the pathogen, they can synthesize quantities of identical "monoclonal" antibodies to link up that target. A monoclonal drug is now being sold to help fight respiratory syncytial virus in babies, and antibodies purified from infected individuals are also used as a treatment for hepatitis B.

Acquired Resistance

Antiviral resistance can be defined by a decreased susceptibility to a drug through either a minimally effective, or completely ineffective, treatment response to prevent associated illnesses from a particular virus. The issue inevitably remains a major obstacle to antiviral therapy as it has developed to almost all specific and effective antimicrobials, including antiviral agents.

The Centers for Disease Control and Prevention (CDC) inclusively recommends those six months and older to get a yearly vaccination to protect from influenza A viruses (H1N1) and (H3N2) and up to two influenza B viruses (depending on the vaccination). Comprehensive protection starts by ensuring vaccinations are current and complete. The three FDA-approved neuraminidase antiviral flu drugs available in the United States, recommended by the CDC, include: oseltamivir (Tamiflu®), zanamivir (Relenza®), and peramivir (Rapivab®).

A study published in 2009 in Nature Biotechnology emphasized the urgent need for augmentation of oseltamivir (Tamiflu®) stockpiles with additional antiviral drugs including zanamivir (Relenza®). This finding was based on a performance evaluation of these drugs supposing the 2009 H1N1 'Swine Flu' neuraminidase (NA) were to acquire the Tamiflu-resistance (His274Tyr) mutation which is currently widespread in seasonal H1N1 strains.

Origin of Antiviral Resistance

The genetic makeup of viruses is constantly changing and therefore may alter the virus resistant to the treatments currently available. Viruses can become resistant through spontaneous or intermittent mechanisms throughout the course of an antiviral treatment. Immunocompromised patients, more often than immunocompetent patients, hospitalized with pneumonia are at the highest risk of developing oseltamivir resistance during treatment. Subsequent to exposure to someone else with the flu, those who received oseltamivir for "post-exposure prophylaxis" are also at higher risk of resistance.

Detection of Antiviral Resistance

National and international surveillance is performed by the CDC to determine effectiveness of the current FDA-approved antiviral flu drugs. Public health officials use this information to make current recommendations about the use of flu antiviral medications. WHO further recommends in-depth epidemiological investigations to control potential transmission of the resistant virus and prevent future progression. As novel treatments and detection techniques to antiviral resistance are enhanced so can the establishment of strategies to combat the inevitable emergence of antiviral resistance.

References

- Wilson, Gisvold; Block, Beale (2004). Wilson and Gisvold's Textbook of Organic Medicinal and Pharmaceutical Chemistry. Philadelphia, Pa.: Lippincott Williams & Wilkins. ISBN 0-7817-3481-9.

- Heymann DL, Aylward RB (2006). "Mass vaccination: when and why". Curr Top Microbiol Immunol. Current Topics in Microbiology and Immunology. 304: 1–16. doi:10.1007/3-540-36583-4_1. ISBN 978-3-540-29382-8. PMID 16989261

- Immunization Safety Review Committee (2004).Immunization Safety Review: Vaccines and Autism. The National Academies Press. ISBN 0-309-09237-X.

- Beringer, Paul; Troy, David A.; Remington, Joseph P. (2006). Remington, the science and practice of pharmacy. Hagerstwon, MD: Lippincott Williams & Wilkins. p. 1419. ISBN 0-7817-4673-6.

- Anderson J, Schiffer C, Lee SK, Swanstrom R (2009). "Viral protease inhibitors". Handb Exp Pharmacol. Handbook of Experimental Pharmacology. 189 (189): 85–110. doi:10.1007/978-3-540-79086-0_4. ISBN 978-3-540-79085-3. PMID 19048198.

- Bai J, Rossi J, Akkina R (March 2001). "Multivalent anti-CCR ribozymes for stem cell-based HIV type 1 gene therapy". AIDS Res. Hum. Retroviruses. 17 (5): 385–99. doi:10.1089/088922201750102427. PMID 11282007.

- "Irradiation of Food FAQ: What is the actual process of irradiation?". U.S. Centers for Disease Control and Prevention. Retrieved 17 April 2016.

- Winegarden, W (June 2014). "The Economics of Pharmaceutical Pricing" (PDF). Pacific Research Institute. Retrieved September 16, 2016.

- Centers for Disease Control and Prevention. Vaccines Do Not Cause Autism. http://www.cdc.gov/vaccinesafety/concerns/autism.html Updated November 23, 2015. Accessed October 20, 2016.

- United Kingdom National Health Service (England). MMR vaccine. http://www.nhs.uk/Conditions/vaccinations/Pages/mmr-vaccine.aspx Last reviewed April 8, 2015. Accessed October 20, 2016.

- "Funding and Guidance for State and Local Health Departments". Centers for Disease Control and Prevention. Retrieved 21 October 2016.

- "Research and Markets: Global Antifungal Therapeutics (Polyenes, Azoles, Echinocandins, Allylamines) Market:Trends and Opportunities (2014-2019) | Business Wire". www.businesswire.com. Retrieved 2015-10-17.

- Wolfgang Saxon (9 June 1999). "Anne Miller, 90, First Patient Who Was Saved by Penicillin". New York Times. Retrieved 29 August 2014.

Antibiotics: An Overview

Antibiotics are drugs used to treat bacteria. With the progress made in antibiotics, diseases such as tuberculosis have been nearly eradicated. Some of the types of antibiotics are clindamycin, dalbavancin, carbapenem, daptomycin etc. The following section will not only provide an overview, it will also delve into the topics related to it.

Antibiotics

Antibiotics, also called antibacterials, are a type of antimicrobial drug used in the treatment and prevention of bacterial infections. They may either kill or inhibit the growth of bacteria. A limited number of antibiotics also possess antiprotozoal activity. Antibiotics are not effective against viruses such as the common cold or influenza, and their inappropriate use allows the emergence of resistant organisms. In 1928, Alexander Fleming identified penicillin, the first chemical compound with antibiotic properties. Fleming was working on a culture of disease-causing bacteria when he noticed the spores of a little green mold (*Penicillium chrysogenum*), in one of his culture plates. He observed that the presence of the mold killed or prevented the growth of the bacteria.

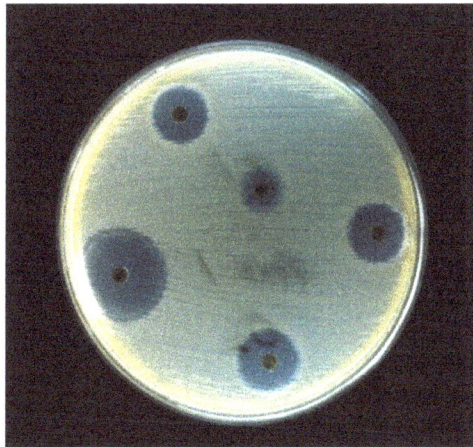

Testing the susceptibility of *Staphylococcus aureus* to antibiotics by the Kirby-Bauer disk diffusion method – antibiotics diffuse from antibiotic-containing disks and inhibit growth of *S. aureus*, resulting in a zone of inhibition.

Antibiotics revolutionized medicine in the 20th century, and have together with vaccination led to the near eradication of diseases such as tuberculosis in the developed world. Their effectiveness and easy access led to overuse, especially in livestock raising, prompting bacteria to develop resistance. This has led to widespread problems with antimicrobial and antibiotic resistance, so much as to prompt the World Health Organization to classify antimicrobial resistance as a "serious threat [that] is no longer a prediction for the future, it is happening right now in every region of the world and has the potential to affect anyone, of any age, in any country".

The era of antibacterial treatment began with the discovery of arsphenamine, first synthesized by Alfred Bertheim and Paul Ehrlich in 1907, and used to treat syphilis. The first systemically active antibacterial drug, prontosil was discovered in 1933 by Gerhard Domagk, for which he was awarded the 1939 Nobel Prize. All classes of antibiotics in use today were first discovered prior to the mid 1980s.

Sometimes the term antibiotic is used to refer to any substance used against microbes, synonymous with antimicrobial, leading to the widespread but incorrect belief that antibiotics can be used against viruses. Some sources distinguish between antibacterial and antibiotic; antibacterials are used in soaps and cleaners generally and antibiotics are used as medicine.

Medical Uses

Antibiotics are used to treat or prevent bacterial infections, and sometimes protozoan infections. (Metronidazole is effective against a number of parasitic diseases). When an infection is suspected of being responsible for an illness but the responsible pathogen has not been identified, an empiric therapy is adopted. This involves the administration of a broad-spectrum antibiotic based on the signs and symptoms presented and is initiated pending laboratory results that can take several days.

When the responsible pathogenic microorganism is already known or has been identified, definitive therapy can be started. This will usually involve the use of a narrow-spectrum antibiotic. The choice of antibiotic given will also be based on its cost. Identification is critically important as it can reduce the cost and toxicity of the antibiotic therapy and also reduce the possibility of the emergence of antimicrobial resistance. To avoid surgery antibiotics may be given for non-complicated acute appendicitis. Effective treatment has been evidenced.

Antibiotics may be given as a preventive measure (prophylactic) and this is usually limited to at-risk populations such as those with a weakened immune system (particularly in HIV cases to prevent pneumonia), those taking immunosuppressive drugs, cancer patients and those having surgery. Their use in surgical procedures is to help prevent infection of incisions made. They have an important role in dental antibiotic prophylaxis where their use may prevent bacteremia and consequent infective endocarditis. Antibiotics are also used to prevent infection in cases of neutropenia particularly cancer-related.

Administration

There are different routes of administration for antibiotic treatment. Antibiotics are usually taken by mouth. In more severe cases, particularly deep-seated systemic infections, antibiotics can be given intravenously or by injection. Where the site of infection is easily accessed antibiotics may be given topically in the form of eye drops onto the conjunctiva for conjunctivitis or ear drops for ear infections and acute cases of swimmer's ear. Topical use is also one of the treatment options for some skin conditions including acne and cellulitis. Advantages of topical application include achieving high and sustained concentration of antibiotic at the site of infection; reducing the potential for systemic absorption and toxicity, and total volumes of antibiotic required are reduced, thereby also reducing the risk of antibiotic misuse. However, some systemic absorption of the antibiotic may occur; the quantity of antibiotic applied is difficult to accurately dose, and there is also the possibility of local hypersensitivity reactions or contact dermatitis occurring.

Side-effects

Health advocacy messages such as this one encourage patients to talk with their doctor about safety in using antibiotics.

Antibiotics are screened for any negative effects before their approval for clinical use, and are usually considered safe and well tolerated. However, some antibiotics have been associated with a wide extent of adverse side effects ranging from mild to very severe depending on the type of antibiotic used, the microbes targeted, and the individual patient. Side effects may reflect the pharmacological or toxicological properties of the antibiotic or may involve hypersensitivity or allergic reactions. Adverse effects range from fever and nausea to major allergic reactions, including photodermatitis and anaphylaxis. Safety profiles of newer drugs are often not as well established as for those that have a long history of use.

Common side-effects include diarrhea, resulting from disruption of the species composition in the intestinal flora, resulting, for example, in overgrowth of pathogenic bacteria, such as *Clostridium difficile*. Antibacterials can also affect the vaginal flora, and may lead to overgrowth of yeast species of the genus *Candida* in the vulvo-vaginal area. Additional side-effects can result from interaction with other drugs, such as the possibility of tendon damage from the administration of a quinolone antibiotic with a systemic corticosteroid.

Obesity

Exposure to antibiotics early in life is associated with increased body mass in humans and mouse models. Early life is a critical period for the establishment of the intestinal microbiota and for metabolic development. Mice exposed to subtherapeutic antibiotic treatment (STAT)– with either penicillin, vancomycin, or chlortetracycline had altered composition of the gut microbiota as well as its metabolic capabilities. One study has reported that mice given low-dose penicillin (1 µg/g body weight) around birth and throughout the weaning process had an increased body mass and fat mass, accelerated growth, and increased hepatic expression of genes involved in adipogenesis, compared to control mice. In addition, penicillin in combination with a high-fat diet increased fasting insulin levels in mice. However, it is unclear whether or not antibiotics cause obesity in humans. Studies have found a correlation between early exposure of antibiotics (<6 months) and increased body mass (at 10 and 20 months). Another study found that the type of antibiotic exposure was also significant with the highest risk of being overweight in those given macrolides compared to penicillin and cephalosporin. Therefore, there is correlation between antibiotic exposure in early life and obesity in humans, but whether or not there is a causal relationship remains unclear. Although there is a correlation between antibiotic use in early life and obesity, the effect of antibiotics on obesity in humans needs to be weighed against the beneficial effects of clinically indicated treatment with antibiotics in infancy.

Interactions

Birth Control Pills

The majority of studies indicate antibiotics do interfere with birth control pills, such as clinical studies that suggest the failure rate of contraceptive pills caused by antibiotics is very low (about 1%). In cases where antibiotics have been suggested to affect the efficiency of birth control pills, such as for the broad-spectrum antibiotic rifampicin, these cases may be due to an increase in the activities of hepatic liver enzymes' causing increased breakdown of the pill's active ingredients. Effects on the intestinal flora, which might result in reduced absorption of estrogens in the colon, have also been suggested, but such suggestions have been inconclusive and controversial. Clinicians have recommended that extra contraceptive measures be applied during therapies using antibiotics that are suspected to interact with oral contraceptives.

Alcohol

Interactions between alcohol and certain antibiotics may occur and may cause side-effects and decreased effectiveness of antibiotic therapy. While moderate alcohol consumption is unlikely to interfere with many common antibiotics, there are specific types of antibiotics with which alcohol consumption may cause serious side-effects. Therefore, potential risks of side-effects and effectiveness depend on the type of antibiotic administered.

Antibiotics such as metronidazole, tinidazole, cephamandole, latamoxef, cefoperazone, cefmenoxime, and furazolidone, cause a disulfiram-like chemical reaction with alcohol by inhibiting its breakdown by acetaldehyde dehydrogenase, which may result in vomiting, nausea, and shortness of breath. In addition, the efficacy of doxycycline and erythromycin succinate may be reduced by alcohol consumption. Other effects of alcohol on antibiotic activity include altered activity of the liver enzymes that break down the antibiotic compound.

Pharmacodynamics

The successful outcome of antimicrobial therapy with antibacterial compounds depends on several factors. These include host defense mechanisms, the location of infection, and the pharmacokinetic and pharmacodynamic properties of the antibacterial. A bactericidal activity of antibacterials may depend on the bacterial growth phase, and it often requires ongoing metabolic activity and division of bacterial cells. These findings are based on laboratory studies, and in clinical settings have also been shown to eliminate bacterial infection. Since the activity of antibacterials depends frequently on its concentration, *in vitro* characterization of antibacterial activity commonly includes the determination of the minimum inhibitory concentration and minimum bactericidal concentration of an antibacterial. To predict clinical outcome, the antimicrobial activity of an antibacterial is usually combined with its pharmacokinetic profile, and several pharmacological parameters are used as markers of drug efficacy.

Combination Therapy

In important infectious diseases, including tuberculosis, combination therapy (i.e., the concurrent application of two or more antibiotics) has been used to delay or prevent the emergence of resistance. In acute bacterial infections, antibiotics as part of combination therapy are prescribed for their synergistic effects to improve treatment outcome as the combined effect of both antibiotics is better than their individual effect. Methicillin-resistant Staphylococcus aureus infections may be treated with a combination therapy of fusidic acid and rifampin. Antibiotics used in combination may also be antagonistic and the combined effects of the two antibiotics may be less than if the individual antibiotic was given as part of a monotherapy. For example, Chloramphenicol and tetracyclines are antagonists to penicillins and aminoglycosides. However, this can vary depending on the species of bacteria. In general, combinations of a bacteriostatic antibiotic and bactericidal antibiotic are antagonistic.

Classes

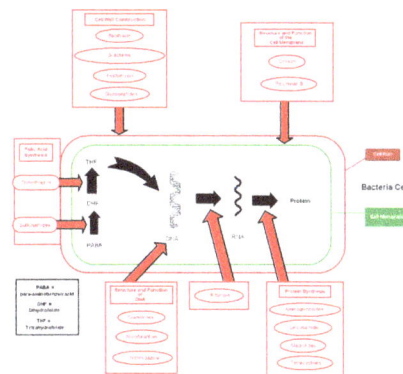

Molecular targets of antibiotics on the bacteria cell

Antibiotics are commonly classified based on their mechanism of action, chemical structure, or spectrum of activity. Most target bacterial functions or growth processes. Those that target the bacterial cell wall (penicillins and cephalosporins) or the cell membrane (polymyxins), or interfere with essential bacterial enzymes (rifamycins, lipiarmycins, quinolones, and sulfonamides) have bactericidal activities. Those that target protein synthesis (macrolides, lincosamides and tetracyclines) are usually bacteriostatic (with the exception of bactericidal aminoglycosides). Further categorization is based on their target specificity. "Narrow-spectrum" antibiotics target specific types of bacteria, such as gram-negative or gram-positive, whereas broad-spectrum antibiotics affect a wide range of bacteria. Following a 40-year break in discovering new classes of antibacterial compounds, four new classes of antibiotics have been brought into clinical use in the late 2000s and early 2010s: cyclic lipopeptides (such as daptomycin), glycylcyclines (such as tigecycline), oxazolidinones (such as linezolid), and lipiarmycins (such as fidaxomicin).

Production

With advances in medicinal chemistry, most modern antibacterials are semisynthetic modifications of various natural compounds. These include, for example, the beta-lactam antibiotics, which include the penicillins (produced by fungi in the genus *Penicillium*), the cephalosporins, and the carbapenems. Compounds that are still isolated from living organisms are the aminoglycosides, whereas other antibacterials—for example, the sulfonamides, the quinolones, and the oxazolidinones—are produced solely by chemical synthesis. Many antibacterial compounds are relatively small molecules with a molecular weight of less than 1000 daltons.

Since the first pioneering efforts of Florey and Chain in 1939, the importance of antibiotics, including antibacterials, to medicine has led to intense research into producing antibacterials at large scales. Following screening of antibacterials against a wide range of bacteria, production of the active compounds is carried out using fermentation, usually in strongly aerobic conditions.

Resistance

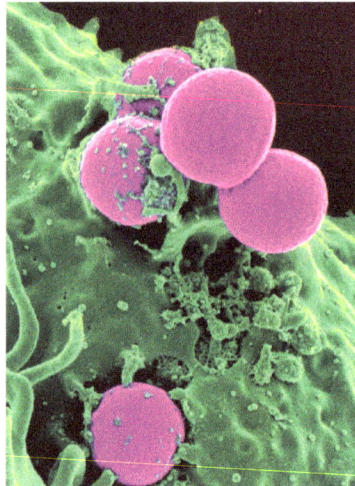

Scanning electron micrograph of a human neutrophil ingesting methicillin-resistant *Staphylococcus aureus* (MRSA)

The emergence of resistance of bacteria to antibiotics is a common phenomenon. Emergence of resistance often reflects evolutionary processes that take place during antibiotic therapy. The antibiotic treatment may select for bacterial strains with physiologically or genetically enhanced capacity to survive high doses of antibiotics. Under certain conditions, it may result in preferential growth of resistant bacteria, while growth of susceptible bacteria is inhibited by the drug. For example, antibacterial selection for strains having previously acquired antibacterial-resistance genes was demonstrated in 1943 by the Luria–Delbrück experiment. Antibiotics such as penicillin and erythromycin, which used to have a high efficacy against many bacterial species and strains, have become less effective, due to the increased resistance of many bacterial strains.

Resistance may take the form of biodegredation of pharmaceuticals, such as sulfamethazine-degrading soil bacteria introduced to sulfamethazine through medicated pig feces. The survival of bacteria often results from an inheritable resistance, but the growth of resistance to antibacterials also occurs through horizontal gene transfer. Horizontal transfer is more likely to happen in locations of frequent antibiotic use.

Antibacterial resistance may impose a biological cost, thereby reducing fitness of resistant strains, which can limit the spread of antibacterial-resistant bacteria, for example, in the absence of antibacterial compounds. Additional mutations, however, may compensate for this fitness cost and can aid the survival of these bacteria.

Paleontological data show that both antibiotics and antibiotic resistance are ancient compounds and mechanisms. Useful antibiotic targets are those for which mutations negatively impact bacterial reproduction or viability.

Several molecular mechanisms of antibacterial resistance exist. Intrinsic antibacterial resistance may be part of the genetic makeup of bacterial strains. For example, an antibiotic target may be absent from the bacterial genome. Acquired resistance results from a mutation in the bacterial chromosome or the acquisition of extra-chromosomal DNA. Antibacterial-producing bacteria have evolved resistance mechanisms that have been shown to be similar to, and may have been transferred to, antibacterial-resistant strains. The spread of antibacterial resistance often occurs through vertical transmission of mutations during growth and by genetic recombination of DNA by horizontal genetic exchange. For instance, antibacterial resistance genes can be exchanged between different bacterial strains or species via plasmids that carry these resistance genes. Plasmids that carry several different resistance genes can confer resistance to multiple antibacterials. Cross-resistance to several antibacterials may also occur when a resistance mechanism encoded by a single gene conveys resistance to more than one antibacterial compound.

Antibacterial-resistant strains and species, sometimes referred to as "superbugs", now contribute to the emergence of diseases that were for a while well controlled. For example, emergent bacterial strains causing tuberculosis (TB) that are resistant to previously effective antibacterial treatments pose many therapeutic challenges. Every year, nearly half a million new cases of multidrug-resistant tuberculosis (MDR-TB) are estimated to occur worldwide. For example, NDM-1 is a newly identified enzyme conveying bacterial resistance to a broad range of beta-lactam antibacterials. The United Kingdom's Health Protection Agency has stated that "most isolates with NDM-1 enzyme are resistant to all standard intravenous antibiotics for treatment of severe infections." On May 26, 2016 an E coli bacteria "superbug" was identified in the United States resistant to colistin, "the last line of defence" antibiotic.

Misuse

Per the *The ICU Book* "The first rule of antibiotics is try not to use them, and the second rule is try not to use too many of them." Inappropriate antibiotic treatment and overuse of antibiotics have contributed to the emergence of antibiotic-resistant bacteria. Self prescription of antibiotics is an example of misuse. Many antibiotics are frequently prescribed to treat symptoms or diseases that do not respond to antibiotics or that are likely to resolve without treatment. Also, incorrect or suboptimal antibiotics are prescribed for certain bacterial infections. The overuse of antibiotics, like penicillin and erythromycin, has been associated with emerging antibiotic resistance since the 1950s. Widespread usage of antibiotics in hospitals has also been associated with increases in bacterial strains and species that no longer respond to treatment with the most common antibiotics.

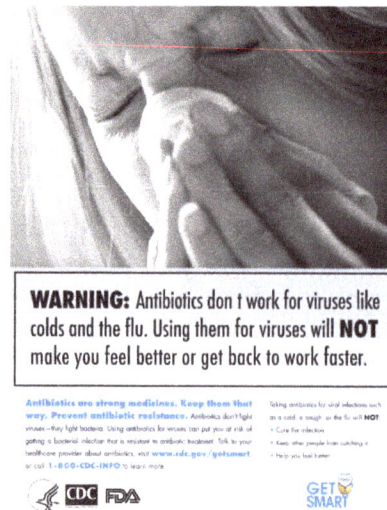

This poster from the US Centers for Disease Control and Prevention "Get Smart" campaign, intended for use in doctors' offices and other healthcare facilities, warns that antibiotics do not work for viral illnesses such as the common cold.

Common forms of antibiotic misuse include excessive use of prophylactic antibiotics in travelers and failure of medical professionals to prescribe the correct dosage of antibiotics on the basis of the patient's weight and history of prior use. Other forms of misuse include failure to take the entire prescribed course of the antibiotic, incorrect dosage and administration, or failure to rest for sufficient recovery. Inappropriate antibiotic treatment, for example, is their prescription to treat viral infections such as the common cold. One study on respiratory tract infections found "physicians were more likely to prescribe antibiotics to patients who appeared to expect them". Multifactorial interventions aimed at both physicians and patients can reduce inappropriate prescription of antibiotics.

Several organizations concerned with antimicrobial resistance are lobbying to eliminate the unnecessary use of antibiotics. The issues of misuse and overuse of antibiotics have been addressed by the formation of the US Interagency Task Force on Antimicrobial Resistance. This task force aims to actively address antimicrobial resistance, and is coordinated by the US Centers for Disease Control and Prevention, the Food and Drug Administration (FDA), and the National Institutes of Health (NIH), as well as other US agencies. An NGO campaign group is *Keep Antibiotics Working*. In France, an "Antibiotics are not automatic" government campaign started in 2002 and led to a marked reduction of unnecessary antibiotic prescriptions, especially in children.

The emergence of antibiotic resistance has prompted restrictions on their use in the UK in 1970 (Swann report 1969), and the EU has banned the use of antibiotics as growth-promotional agents since 2003. Moreover, several organizations (including the World Health Organization, the National Academy of Sciences, and the U.S. Food and Drug Administration) have advocated restricting the amount of antibiotic use in food animal production. However, commonly there are delays in regulatory and legislative actions to limit the use of antibiotics, attributable partly to resistance against such regulation by industries using or selling antibiotics, and to the time required for research to test causal links between their use and resistance to them. Two federal bills (S.742 and H.R. 2562) aimed at phasing out nontherapeutic use of antibiotics in US food animals were proposed, but have not passed. These bills were endorsed by public health and medical organizations, including the American Holistic Nurses' Association, the American Medical Association, and the American Public Health Association (APHA).

There has been extensive use of antibiotics in animal husbandry. In the United States, the question of emergence of antibiotic-resistant bacterial strains due to use of antibiotics in livestock was raised by the US Food and Drug Administration (FDA) in 1977. In March 2012, the United States District Court for the Southern District of New York, ruling in an action brought by the Natural Resources Defense Council and others, ordered the FDA to revoke approvals for the use of antibiotics in livestock, which violated FDA regulations.

History

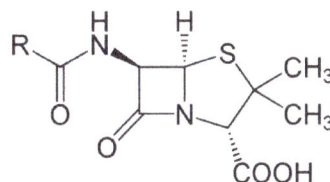

Penicillin, the first natural antibiotic discovered by Alexander Fleming in 1928

Before the early 20th century, treatments for infections were based primarily on medicinal folklore. Mixtures with antimicrobial properties that were used in treatments of infections were described over 2000 years ago. Many ancient cultures, including the ancient Egyptians and ancient Greeks, used specially selected mold and plant materials and extracts to treat infections. More recent observations made in the laboratory of antibiosis between microorganisms led to the discovery of natural antibacterials produced by microorganisms. Louis Pasteur observed, "if we could intervene in the antagonism observed between some bacteria, it would offer perhaps the greatest hopes for therapeutics". The term 'antibiosis', meaning "against life", was introduced by the French bacteriologist Jean Paul Vuillemin as a descriptive name of the phenomenon exhibited by these early antibacterial drugs. Antibiosis was first described in 1877 in bacteria when Louis Pasteur and Robert Koch observed that an airborne bacillus could inhibit the growth of *Bacillus anthracis*. These drugs were later renamed antibiotics by Selman Waksman, an American microbiologist, in 1942. Synthetic antibiotic chemotherapy as a science and development of antibacterials began in Germany with Paul Ehrlich in the late 1880s. Ehrlich noted certain dyes would color human, animal, or bacterial cells, whereas others did not. He then proposed the idea that it might be possible to create chemicals that would act as a selective drug that would bind to and kill bacteria without harming the human host. After screening hundreds of dyes against various organisms, in 1907, he discovered a medicinally useful drug, the first synthetic antibacterial salvarsan now called arsphenamine.

The effects of some types of mold on infection had been noticed many times over the course of history. In 1928, Alexander Fleming noticed the same effect in a Petri dish, where a number of disease-causing bacteria were killed by a fungus of the genus *Penicillium*. Fleming postulated that the effect is mediated by an antibacterial compound he named penicillin, and that its antibacterial properties could be exploited for chemotherapy. He initially characterized some of its biological properties, and attempted to use a crude preparation to treat some infections, but he was unable to pursue its further development without the aid of trained chemists.

Alexander Fleming

The first sulfonamide and first commercially available antibacterial, Prontosil, was developed by a research team led by Gerhard Domagk in 1932 at the Bayer Laboratories of the IG Farben conglomerate in Germany. Domagk received the 1939 Nobel Prize for Medicine for his efforts. Prontosil had a relatively broad effect against Gram-positive cocci, but not against enterobacteria. Research was stimulated apace by its success. The discovery and development of this sulfonamide drug opened the era of antibacterials.

In 1939, coinciding with the start of World War II, Rene Dubos reported the discovery of the first naturally derived antibiotic, tyrothricin, a compound of 20% gramicidin and 80% tyrocidine, from *B. brevis*. It was one of the first commercially manufactured antibiotics universally and was very effective in treating wounds and ulcers during World War II. Gramicidin, however, could not be used systemically because of toxicity. Tyrocidine also proved too toxic for systemic usage. Research results obtained during that period were not shared between the Axis and the Allied powers during World War II and limited access during the Cold War.

Florey and Chain succeeded in purifying the first penicillin, penicillin G, in 1942, but it did not become widely available outside the Allied military before 1945. Later, Norman Heatley developed the back extraction technique for efficiently purifying penicillin in bulk. The chemical structure of penicillin was determined by Dorothy Crowfoot Hodgkin in 1945. Purified penicillin displayed potent antibacterial activity against a wide range of bacteria and had low toxicity in humans. Furthermore, its activity was not inhibited by biological constituents such as pus, unlike the synthetic sulfonamides. The discovery of such a powerful antibiotic was unprecedented, and the development of penicillin led to renewed interest in the search for antibiotic compounds with similar efficacy and safety. For their successful development of penicillin, which Fleming had accidentally discovered but could not develop himself, as a therapeutic drug, Ernst Chain and Howard Florey shared the 1945 Nobel Prize in Medicine with Fleming. Florey credited Dubos with pioneering the approach of deliberately and systematically searching for antibacterial compounds, which had led to the discovery of gramicidin and had revived Florey's research in penicillin.

Etymology

The term *antibiotic* was first used in 1942 by Selman Waksman and his collaborators in journal articles to describe any substance produced by a microorganism that is antagonistic to the growth of other microorganisms in high dilution. This definition excluded substances that kill bacteria but that are not produced by microorganisms (such as gastric juices and hydrogen peroxide). It also excluded synthetic antibacterial compounds such as the sulfonamides. In current usage, the term "antibiotic" is applied to any medication that kills bacteria or inhibits their growth, regardless of whether that medication is produced by a microorganism or not.

Research

Alternatives

The increase in bacterial strains that are resistant to conventional antibacterial therapies has prompted the development of bacterial disease treatment strategies that are alternatives to conventional antibacterials, including phage therapy.

Resistance-modifying Agents

One strategy to address bacterial drug resistance is the discovery and application of compounds that modify resistance to common antibacterials. Resistance modifying agents are capable of partly or completely suppressing bacterial resistance mechanisms. For example, some resistance-modifying agents may inhibit multidrug resistance mechanisms, such as drug efflux from the cell, thus increasing the susceptibility of bacteria to an antibacterial. Targets include:

- The efflux inhibitor Phe-Arg-β-naphthylamide.
- Beta-lactamase inhibitors, such as clavulanic acid and sulbactam

Metabolic stimuli such as sugar can help eradicate a certain type of antibiotic-tolerant bacteria by keeping their metabolism active.

Vaccines

Vaccines rely on immune modulation or augmentation. Vaccination either excites or reinforces the immune competence of a host to ward off infection, leading to the activation of macrophages, the production of antibodies, inflammation, and other classic immune reactions. Antibacterial vaccines have been responsible for a drastic reduction in global bacterial diseases. Vaccines made from attenuated whole cells or lysates have been replaced largely by less reactogenic, cell-free vaccines consisting of purified components, including capsular polysaccharides and their conjugates, to protein carriers, as well as inactivated toxins (toxoids) and proteins.

Phage Therapy

Phage therapy is another method for treating antibiotic-resistant strains of bacteria. Phage therapy infects pathogenic bacteria with their own viruses, bacteriophages. Bacteriophages, also known simply as phages, infect and can kill bacteria. Phages insert their DNA into the bacterium, where it is transcribed and used to make new phages, after which the cell will lyse, releasing new phage able to infect and destroy further bacteria of the same strain. The high specificity of phage protects "good" bacteria from destruction. When applicable, bacteriophage therapy defeats antibiotic resistant bacteria.

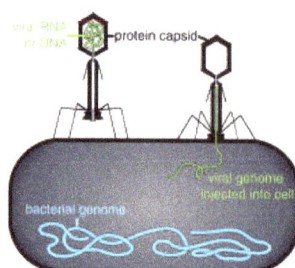

Phage injecting its genome into bacterial cell

Supplements

Some antioxidant dietary supplements contain polyphenols, such as grape seed extract, and demonstrate *in vitro* anti-bacterial properties.

Development of New Antibiotics

In April 2013, the Infectious Disease Society of America (IDSA) reported that the weak antibiotic pipeline does not match bacteria's increasing ability to develop resistance. Since 2009, only 2 new antibiotics were approved in the United States. The number of new antibiotics approved for marketing per year declines continuously. The report identified seven antibiotics against the Gram-negative bacilli (GNB) currently in phase 2 or phase 3 clinical trials. However, these drugs do not address the entire spectrum of resistance of GNB. Some of these antibiotics are combination of existent treatments:

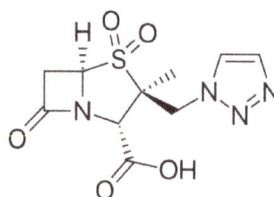

Tazobactam

- Ceftolozane/tazobactam (CXA-201; CXA-101/tazobactam): Antipseudomonal cephalosporin/β-lactamase inhibitor combination (cell wall synthesis inhibitor). FDA approved on 12/19/2014.

- Ceftazidime/avibactam (ceftazidime/NXL104): Antipseudomonal cephalosporin/β-lactamase inhibitor combination (cell wall synthesis inhibitor). In phase 3.

- Ceftaroline/avibactam (CPT-avibactam; ceftaroline/NXL104): Anti-MRSA cephalosporin/β-lactamase inhibitor combination (cell wall synthesis inhibitor)

- Imipenem/MK-7655: Carbapenem/ β-lactamase inhibitor combination (cell wall synthesis inhibitor). In phase 2.

- Plazomicin (ACHN-490): Aminoglycoside (protein synthesis inhibitor). In phase 2.

- Eravacycline (TP-434): Synthetic tetracycline derivative / protein synthesis inhibitor targeting the ribosome. Development by Tetraphase, Phase 2 trials complete.

- Brilacidin (PMX-30063): Peptide defense protein mimetic (cell membrane disruption). In phase 2.

Streptomyces research is expected to provide new antibiotics, including treatment against MRSA and infections resistant to commonly used medication. Efforts of John Innes Centre and universities in the UK, supported by BBSRC, resulted in the creation of spin-out companies, for example Novacta Biosystems, which has designed the type-b lantibiotic-based compound NVB302 (in phase 1) to treat *Clostridium difficile* infections. Possible improvements include clarification of clinical trial regulations by FDA. Furthermore, appropriate economic incentives could persuade pharmaceutical companies to invest in this endeavor. In the US, the Antibiotic Development to Advance Patient Treatment (ADAPT) Act was introduced with the aim of fast tracking the drug development of antibiotics to combat the growing threat of 'superbugs'. Under this Act, FDA can approve antibiotics and antifungals treating life-threatening infections based on smaller clinical trials. The CDC will monitor the use of antibiotics and the emerging resistance, and publish the data. The FDA antibiotics labeling process, 'Susceptibility Test Interpretive Criteria for Microbial Organisms' or 'breakpoints', will provide accurate data to healthcare professionals. According to Allan Coukell, senior director for health programs at The Pew Charitable Trusts, "By allowing drug developers to rely on smaller datasets, and clarifying FDA's authority to tolerate a higher level of uncertainty for these drugs when making a risk/benefit calculation, ADAPT would make the clinical trials more feasible."

Types of Antibiotics

Clindamycin

Clindamycin is an antibiotic useful for the treatment of a number of bacterial infections. This includes middle ear infections, bone or joint infections, pelvic inflammatory disease, strep throat, pneumonia, and endocarditis among others. It can be useful against some cases of methicillin-resistant *Staphylococcus aureus* (MRSA). It may also be used for acne and in addition to quinine for malaria. It is available by mouth, intravenously, and as a cream to be applied to the skin or in the vagina.

Common side effects include nausea, diarrhea, rash, and pain at the site of injection. It increases the risk of *Clostridium difficile* colitis about fourfold. Other antibiotics may be recommended instead due to this reason. It appears to be generally safe in pregnancy. It is of the lincosamide class and works by blocking bacteria from making protein.

Clindamycin was first made in 1967. It is on the World Health Organization's List of Essential Medicines, the most important medication needed in a basic health system. It is available as a generic medication and is not very expensive. The wholesale cost in the developing world is about 0.06 to 0.12 USD per pill. In the United States it costs about 2.70 USD a dose.

Medical Uses

Clindamycin is used primarily to treat anaerobic infections caused by susceptible anaerobic bacteria, including dental infections, and infections of the respiratory tract, skin, and soft tissue, and peritonitis. In people with hypersensitivity to penicillins, clindamycin may be used to treat infections caused by susceptible aerobic bacteria, as well. It is also used to treat bone and joint infections, particularly those caused by *Staphylococcus aureus*. Topical application of clindamycin phosphate can be used to treat mild to moderate acne.

Acne

The use of clindamycin in conjunction with benzoyl peroxide is more effective in the treatment of acne than the use of either product by itself.

Clindamycin and adapalene in combination are also more effective than either drug alone, although adverse effects are more frequent.

Susceptible Bacteria

It is most effective against infections involving the following types of organisms:

- Aerobic Gram-positive cocci, including some members of the *Staphylococcus* and *Streptococcus* (e.g. pneumococcus) genera, but not enterococci.

- Anaerobic, Gram-negative rod-shaped bacteria, including some *Bacteroides*, *Fusobacterium*, and *Prevotella*, although resistance is increasing in *Bacteroides fragilis*.

Most aerobic Gram-negative bacteria (such as *Pseudomonas*, *Legionella*, *Haemophilus influenzae* and *Moraxella*) are resistant to clindamycin, as are the facultative anaerobic Enterobacteriaceae. A notable exception is *Capnocytophaga canimorsus*, for which clindamycin is a first-line drug of choice.

The following represents MIC susceptibility data for a few medically significant pathogens.

- *Staphylococcus aureus*: 0.016 µg/ml - >256 µg/ml

- *Streptococcus pneumoniae*: 0.002 µg/ml - >256 µg/ml

- *Streptococcus pyogenes*: <0.015 µg/ml - >64 µg/ml

D-Test

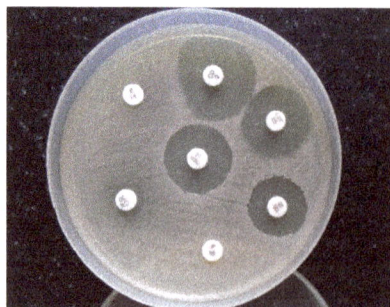

D test

When testing a Gram-positive culture for sensitivity to clindamycin, it is common to perform a "D-Test" to determine if there is a macrolide-resistant subpopulation of bacteria present. This test is necessary because some bacteria express a phenotype known as MLS_B, in which susceptibility tests will indicate the bacteria are susceptible to clindamycin, but *in vitro* the pathogen displays inducible resistance.

To perform a D-test, an agar plate is inoculated with the bacteria in question and two drug-impregnated disks (one with erythromycin, one with clindamycin) are placed 15–20 mm apart on the plate. If the area of inhibition around the clindamycin disk is "D" shaped, the test result is positive and clindamycin should not be used due to the possibility of resistant pathogens and therapy failure. If the area of inhibition around the clindamycin disk is circular, the test result is negative and clindamycin can be used.

Malaria

Given with chloroquine or quinine, clindamycin is effective and well tolerated in treating *Plasmodium falciparum* malaria; the latter combination is particularly useful for children, and is the treatment of choice for pregnant women who become infected in areas where resistance to chloroquine is common. Clindamycin should not be used as an antimalarial by itself, although it appears to be very effective as such, because of its slow action. Patient-derived isolates of *Plasmodium falciparum* from the Peruvian Amazon have been reported to be resistant to clindamycin as evidenced by *in vitro* drug susceptibility testing.

Other

Clindamycin may be useful in skin and soft tissue infections caused by methicillin-resistant *Staphylococcus aureus* (MRSA); many strains of MRSA are still susceptible to clindamycin; however, in the United States spreading from the West Coast eastwards, MRSA is becoming increasingly resistant.

Clindamycin is used in cases of suspected toxic shock syndrome, often in combination with a bacteriostatic agent such as vancomycin. The rationale for this approach is a presumed synergy between vancomycin, which causes the death of the bacteria by breakdown of the cell wall, and clindamycin, which is a powerful inhibitor of toxin synthesis. Both *in vitro* and *in vivo* studies have shown clindamycin reduces the production of exotoxins by staphylococci; it may also induce changes in the surface structure of bacteria that make them more sensitive to immune system attack (opsonization and phagocytosis).

Clindamycin has been proven to decrease the risk of premature births in women diagnosed with bacterial vaginosis during early pregnancy to about a third of the risk of untreated women.

The combination of clindamycin and quinine is the standard treatment for severe babesiosis.

Clindamycin may also be used to treat toxoplasmosis, and, in combination with primaquine, is effective in treating mild to moderate *Pneumocystis jirovecii* pneumonia.

Side Effects

Common adverse drug reactions associated with systemic clindamycin therapy — found in over 1% of people — include: diarrhea, pseudomembranous colitis, nausea, vomiting, abdominal pain or cramps and/or rash. High doses (both intravenous and oral) may cause a metallic taste. Common adverse drug reactions associated with topical formulations - found in over 10% of people - include: dryness, burning, itching, scaliness, or peeling of skin (lotion, solution); erythema (foam, lotion, solution); oiliness (gel, lotion). Additional side effects include contact dermatitis. Common side effects - found in over 10% of people - in vaginal applications include fungal infection.

Pseudomembranous colitis is a potentially lethal condition commonly associated with clindamycin, but which occurs with other antibiotics, as well. Overgrowth of *Clostridium difficile*, which is inherently resistant to clindamycin, results in the production of a toxin that causes a range of adverse effects, from diarrhea to colitis and toxic megacolon.

Rarely — in less than 0.1% of patients — clindamycin therapy has been associated with anaphylaxis, blood dyscrasias, polyarthritis, jaundice, raised liver enzyme levels, renal dysfunction, cardiac arrest, and/or hepatotoxicity.

Interactions

Clindamycin may prolong the effects of neuromuscular-blocking drugs, such as succinylcholine and vecuronium. Its similarity to the mechanism of action of macrolides and chloramphenicol means they should not be given simultaneously, as this causes antagonism and possible cross-resistance.

Chemistry

Clindamycin is a semisynthetic derivative of lincomycin, a natural antibiotic produced by the actinobacterium *Streptomyces lincolnensis*. It is obtained by 7(*S*)-chloro-substitution of the 7(*R*)-hydroxyl group of lincomycin. The synthesis of clindamycin was first announced by BJ Magerlein, RD Birkenmeyer, and F Kagan on the fifth Interscience Conference on Antimicrobial Agents and Chemotherapy (ICAAC) in 1966. It has been on the market since 1968.

Mechanism of Action

Clindamycin has a primarily bacteriostatic effect. It is a bacterial protein synthesis inhibitor by inhibiting ribosomal translocation, in a similar way to macrolides. It does so by binding to the 50S rRNA of the large bacterial ribosome subunit.

The structures of the complexes between several antibiotics (including clindamycin) and a *Deinococcus radiodurans* ribosome have been solved by X-ray crystallography by a team from the Max Planck Working Groups for Structural Molecular Biology, and published in the journal *Nature*.

Society and Culture

Cost

It is available as a generic medication and is not very expensive. The wholesale cost in the developing world is about 0.06 to 0.12 USD per pill. In the United States it costs about 2.70 USD a dose.

The wholesale price in UK is less than 5 pence per pill, however the RX system covers the cost for citizens. Canada and Mexico also have a similar cost, with average price of 4 cents per pill.

Available Forms

Clindamycin preparations for oral administration include capsules (containing clindamycin hydrochloride) and oral suspensions (containing clindamycin palmitate hydrochloride). Oral suspension is not favored for administration of clindamycin to children, due to its extremely foul taste and odor. Clindamycin is formulated in a vaginal cream and as vaginal ovules for treatment of bacterial vaginosis. It is also available for topical administration in gel form, as a lotion, and in a foam delivery system (each containing clindamycin phosphate) and a solution in ethanol (containing clindamycin hydrochloride) and is used primarily as a prescription acne treatment.

Several combination acne treatments containing clindamycin are also marketed, such as single-product formulations of clindamycin with benzoyl peroxide—sold as BenzaClin (Sanofi-Aventis), Duac (a gel form made by Stiefel), and Acanya, among other trade names—and, in the United States, a combination of clindamycin and tretinoin, sold as Ziana. In India, vaginal suppositories containing clindamycin in combination with clotrimazole are manufactured by Olive Health Care and sold as Clinsup-V. In Egypt, vaginal cream containing clindamycin produced by Biopharmgroup sold as Vagiclind indicated for vaginosis.

Clindamycin is available as a generic drug, for both systemic (oral and intravenous) and topical use (The exception is the vaginal suppository, which is not available as a generic in the USA).

Clindamycin is marketed as generic and under trade names including Cleocin HCl, Dalacin, Lincocin (Bangladesh), Dalacin, and Clindacin. Combination products include Duac, BenzaClin, Clindoxyl and Acanya (in combination with benzoyl peroxide), and Ziana (with tretinoin).

Veterinary Use

The veterinary uses of clindamycin are quite similar to its human indications, and include treatment of osteomyelitis, skin infections, and toxoplasmosis, for which it is the preferred drug in dogs and cats. Toxoplasmosis rarely causes symptoms in cats, but can do so in very young or immunocompromised kittens and cats.

Dalbavancin

Dalbavancin (INN, trade names Dalvance in the US and Xydalba in Europe) is a novel second-generation lipoglycopeptide antibiotic. It belongs to the same class as vancomycin, the most widely used and one of the few treatments available to patients infected with methicillin-resistant *Staphylococcus aureus* (MRSA).

Dalbavancin is a semisynthetic lipoglycopeptide that was designed to improve upon the natural glycopeptides currently available, vancomycin and teicoplanin.

It possesses *in vitro* activity against a variety of Gram-positive pathogens including MRSA and methicillin-resistant *Staphylococcus epidermidis* (MRSE). It is a once-weekly, two-dose antibiotic, the rights to which Actavis acquired when it bought Durata Therapeutics in 2014.

The Food and Drug Administration approved dalbavancin in May 2014 for the treatment of acute bacterial skin and skin structure infections (ABSSSIs) caused by certain susceptible bacteria such as *Staphylococcus aureus* including methicillin-susceptible and methicillin-resistant strains of *Streptococcus pyogenes*, in intravenous dosage form.

Medical Uses

Dalbavancin is an antibiotic used to treat acute bacterial skin and skin structure infections (ABSSSI) in adults caused by susceptible Gram-positive organisms, including methicillin-resistant *Staphylococcus aureus* (MRSA). MRSA infections have become problematic in the community and in healthcare settings due to resistance to many available antibiotics. Because dalbavancin has demonstrated efficacy against MRSA and other microorganisms to treat serious or life-threatening infections, it was the first drug approved as a Qualified Infectious Disease Product under the Generating Antibiotic Incentives Now (GAIN) act, which is part of the FDA Safety and Innovation Act.

It has strong activity against many Gram-positive bacteria, including methicillin-sensitive and methicillin-resistant *Staphylococcus aureus*, *Streptococcus pyogenes*, *Streptococcus agalactiae*, *Streptococcus anginosus*, *Streptococcus intermedius*, and *Streptococcus constellatus*. Based on MIC data and other studies, dalbavancin is more potent and bactericidal and therefore requires lower concentrations than vancomycin against these organisms. Dalbavancin also shows *in vitro* activity against vancomycin-susceptible *Enterococcus faecium* and *Enterococcus faecalis*. Other Gram-positive organisms belonging to the *Bacillus* spp., *Listeria* spp., and *Corynebacterium* spp. may show *in vitro* susceptibility, and dalbavancin may exhibit activity against enterococci expressing the VanB or VanC phenotype of acquired resistance against vancomycin. There is no clinically significant activity against Gram-negative bacteria.

Two trials, DISCOVER 1 and DISCOVER 2, demonstrated noninferiority of dalbavancin compared to vancomycin/linezolid in the treatment of ABSSSI. Patients were randomly assigned to receive two doses of dalbavancin on days 1 and 8, or to receive intravenous vancomyin for at least 3 days with the option of switching to oral linezolid to complete 10 to 14 days of therapy. Investigators assessed the cessation of spread of infection-related erythema and the absence of fever at 48 to 72 hours as the primary endpoint and found that once-weekly dalbavancin was as effective as twice-daily intravenous vancomycin followed by oral linezolid. This once-weekly dosing regimen may offer an advantageous treatment option compared to daily or twice-daily dosing.

Contraindications

Dalbavancin is contraindicated in patients with hypersensitivity to dalbavancin, such as skin reactions or anaphylaxis, and caution is advised for patients with known hypersensitivity to other glycopeptides. There is currently no data on cross-reactivity between dalbavancin and vancomyin.

Side Effects

The most common adverse reactions encountered in Phase II and Phase III trials were nausea (5.5%), headache (4.7%), and diarrhea (4.4%), as well as rash (2.7%) and itchiness (2.1%). Other less frequent but serious adverse reactions included hematologic disorders, hepatotoxicity, *Clostridium difficile* colitis, bronchospasm, infusion-related reactions including Red Man Syndrome, and anaphylactic shock. In trials, dalbavancin was associated with higher rates of hemorrhagic events compared to comparator groups and should be a precaution in patients undergoing surgery or taking anticoagulants. Patients on dalbavancin also had post-baseline alanine aminotransferase (ALT) levels that were 3 times the upper normal limit, some even having elevations 10 times the upper normal limit; however, eight of the twelve dalbavancin-treated patients had comorbid conditions that could affect their ALT, compared to only one patient in the comparator group. There is no evidence of ototoxicity associated with dalbavancin.

Drug Interactions

Clinical drug-drug interactions with dalbavancin have not been studied, and dalbavancin does not appear to interact with cytochrome P450 substrates, inhibitors, or inducers. It was found to have an *in vitro* synergistic interaction with the antimicrobial oxacillin, but the clinical significance of this interaction has yet to be established.

Pregnancy and Lactation

Use of dalbavancin in pregnant women has not been studied sufficiently and should only occur when the potential benefit outweighs the potential risk to the fetus. Animal studies did not show embryo or fetal toxicity at doses that were 1.2 and 0.7 times the human dose. However, delayed fetal maturation was observed at a dose that was 3.5 times the human dose. While dalbavancin is excreted in rat milk, it is unknown if it is excreted in human milk. It should be used in nursing mothers only when the potential benefit exceeds the potential risk. There is no evidence in animals of teratogenicity.

Mechanism of Action

Dalbavancin is a lipoglycopeptide belonging in the same glycopeptide class as vancomycin. Similar to other glycopeptides, dalbavancin exerts its bactericidal effect by disrupting cell wall biosynthesis. It binds to the D-alanyl-D-alanyl residue on growing peptidoglycan chains and prevents transpeptidation from occurring, preventing peptidoglycan elongation and cell wall formation. Dalbavancin also dimerizes and anchors itself in the lipophilic bacterial membrane, thereby increasing its stability in the target environment and its affinity for peptidoglycan.

Antimicrobial activity correlates with the ratio of area under the concentration-time curve to minimum inhibitory concentration for *Staphylococcus aureus*.

History

Dalbavancin has undergone a phase-III clinical trial for adults with complicated skin infections, but in December 2007, the FDA said more data were needed before approval. On September 9, 2008, Pfizer announced it will withdraw all marketing applications to conduct another phase-III clinical trial. Durata Therapeutics acquired the rights to dalbavancin in December 2009, and has initiated two new phase-III clinical trials for treatment of ABSSSIs. Preliminary results in 2012 were promising.

About 1,289 adults with ABSSSI were given dalbavancin or vancomycin randomly, and dalbavancin was found to exhibit efficacy comparable to vancomycin.

In May 2014, dalbavancin was approved by the FDA for use in the US for ABSSSIs, including MRSA and *Streptococcus pyogenes* infections.

Carbapenem

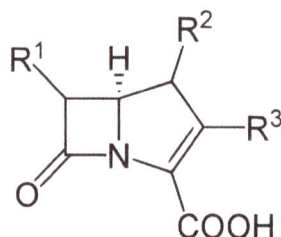

Core structure of the carbapenem molecules

Carbapenems are antibiotics used for the treatment of infections known or suspected to be caused by multidrug-resistant (MDR) bacteria. Their use is primarily in people who are hospitalized.

Like the penicillins and cephalosporins, they are members of the beta lactam class of antibiotics, which kill bacteria by binding to penicillin-binding proteins and inhibiting cell wall synthesis. They exhibit a broader spectrum of activity compared to cephalosporins and penicillins. Their effectiveness is less affected by many common mechanisms of antibiotic resistance than other beta lactams.

Carbapenem antibiotics were originally developed at Merck & Co. from the carbapenem thienamycin, a naturally derived product of *Streptomyces cattleya*. Concern has arisen in recent years over increasing rates of resistance to carbapenems, as there are few therapeutic options for treating infections caused by carbapenem-resistant bacteria (such as the carbapenem-resistant Enterobacteriaceae and Klebsiella pneumoniae).

Medical Uses

Abdominal Infections

The carbapenem ertapenem is one of several first-line agents recommended by the Infectious Disease Society of America for the empiric treatment of community-acquired intra-abdominal infections of mild-to-moderate severity. Agents with anti-pseudomonal activity, including doripenem, imipenem, and meropenem are not recommended in this population. Doripenem, imipenem, and meropenem are recommended for high-risk community-acquired abdominal infections and for abdominal infections that are hospital-acquired.

Complicated Urinary Tract Infections

A 2015 systematic review found little evidence that would support the identification of a best antimicrobial regimen for complicated urinary tract infections, but identified three high-quality trials supporting high cure rates with doripenem, including in patients with levofloxacin-resistant *E. coli* infections.

Pneumonia

The carbapenems imipenem and meropenem are recommended by the American Thoracic Society and the Infectious Disease Society of America as one of several first-line therapy options for people with late-onset hospital-acquired or ventilator-associated pneumonia, especially when *Pseudomonas*, *Acinetobacter*, or extended spectrum beta-lactamase producing *Enterobactericeace* are suspected pathogens. Combination therapy, typically with an aminoglycoside, is recommended for *Pseudomonas* infections to avoid resistance development during treatment.

Carbapenems are less commonly used in the treatment of community-acquired pneumonia, as community-acquired strains of the most common responsible pathogens (*Streptocuccus pneumoniae*, *Haemophilus influenazae*, atypical bacteria, and Enterobactericeace) are typically susceptible to narrower spectrum and/or orally administered agents such as fluoroquinolones, amoxicillin, or azithromycin. Imipenem and meropenem are useful in cases in which *P. aeruginosa* is a suspected pathogen.

Bloodstream Infections

A 2015 meta analysis concluded that the anti-pseudomonal penicillin-beta lactamase inhibitor combination piperacillin-tazobactam gives results equivalent to treatment with a carbapenem in patients with sepsis. In 2015, the National Institute for Health and Care Excellence recommended piperacillin-tazobactam as first line therapy for the treatment of bloodstream infections in neutropenic cancer patients.

For bloodstream infections known to be due to extended spectrum beta-lactamase producing *Enterobacteriaceace*, carbapenems are superior to alternative treatments.

Spectrum of Activity

Carbapenems exhibit broad spectrum activity against gram-negative bacteria and somewhat narrower activity against gram-positive bacteria. For empiric therapy (treatment of infections prior to identification of the responsible pathogen) they are often combined with a second drug having broader spectrum gram-positive activity.

Gram-negative Pathogens

The spectrum of activity of the carbapenems imipenem, doripenem, and meropenem includes most *Enterobacteriaceace* species, including *Escherichia coli*, *Klebsiella pneumoniae*, *Enterobacter cloacae*, *Citrobacter freundii*, *Proteus mirabilis*, and *Serratia marcescens*. Activity is maintained against most strains of *E. coli* and *K. pneumoniae* that are resistant to cephalosporins due to the production of extended spectrum beta-lactamases. Imipenem, doripenem, and meropenem also exhibit good activity against most strains of *Pseudomonas aeruginosa* and *Acinetobacter* species. The observed activity against these pathogens is especially valued as they are intrinsically resistant to many other antibiotic classes.

Gram-positive Pathogens

The spectrum of activity of the carbapenems against gram-positive bacteria is fairly broad, but not as exceptionally so as in the case of gram-negative bacteria. Good activity is seen against methicillin-sensitive strains of *Staphylococcus* species, but many other antibiotics provide coverage for such infections. Good activity is also observed for most *Streptococcus* species, including penicillin-resistant strains. Carbapenems are not highly active against methicillin-resistant *Staphyloccus aureus* or most *enterococcal* infections because carbapenems do not bind to the penicillin-binding protein used by these pathogens.

Other

Carbapenems generally exhibit good activity against anaerobes such as *Bacteriodes fragilis*. Like other beta lactam antibiotics, they lack activity against atypical bacteria, which do not have a cell wall and are thus not affected by cell wall synthesis inhibitors.

Contraindications

Carbapenems are contraindicated in patients with prior allergic reactions to beta lactam antibiotics. In addition, as the intramuscular formulations of ertrapenem and imipenem are formulated with lidocaine, the intramuscular formulation of these two drugs are contraindicated in patients with prior adverse reactions to lidocaine.

Adverse Effects

Serious and occasionally fatal allergic reactions can occur in people treated with carbapenems. Seizures are a dose-limiting toxicity for both imipenem and meropenem. *Clostridium difficile*-related diarrhea may occur in people treated with carbapenems or other broad spectrum antibiotics.

Examples

Approved for Clinical Use

Imipenem, the first clinically used carbapenem, was developed at Merck and Co. It was approved for use in the United States in 1985. Imipenem is hydrolyzed in the mammalian kidney by a dehydropeptidase enzyme to a nephrotoxic intermediate, and thus is co-formulated with the dehydropeptidase inhibitor cilastatin. Imipenem is available in both intravenous and intramuscular formulations.

Meropenem is stable to mammalian dehydropeptidases and does not require co-administration of cilastin. It was approved for use in the United States in 1996. In most indications it is somewhat more convenient to administer than imipenem, 3 times a day rather than 4. Doses of less than one gram may be administered as an IV bolus, whereas imipenem is usually administered as a 20-minute to one hour infusion. Meropenem is somewhat less potent than imipenem against gram-positive pathogens, and somewhat more potent against gram-negative infections. Unlike imipenem, which produced an unacceptable rate of seizures in a phase 2 trial, meropenem is effective for the treatment of bacterial meningitis. A systematic review performed by an employee of the company that markets meropenem concluded that it provides a higher bacterial response and lower adverse event rates than imipenem in people with severe infections, but no difference in mortality rate.

Ertapenem is administered once daily as an intravenous infusion or intramuscular injection. It lacks useful activity against the *P. aeruginosa* and *Acetinobacter* species, both of which are important causes of hospital-acquired infections.

Doripenem has a spectrum of activity very similar to that of meropenem. Its greater stability in solution allows the use of prolonged infusions and it is somewhat less likely to produce seizures than other carbapenems.

Panipenem/betamipron (Japanese approval 1993)

Biapenem (Japanese approval 2001) exhibits similar efficacy and adverse event rates as other carbapenems.

Unapproved/Experimental

- Razupenem (PZ-601)

    - PZ-601 is a carbapenem antibiotic currently being tested as having a broad spectrum of activity including strains resistant to other carbapenems. Despite early Phase II promise, Novartis (who acquired PZ-601 in a merger deal with Protez Pharmaceuticals) recently dropped PZ-601, citing a high rate of adverse events in testing.

- Tebipenem

    - Tebipenem is the first carbapenem whose prodrug form, the pivalyl ester, is orally available.

- Lenapenem

- Tomopenem

- Thienamycin (thienpenem) the first discovered carbapenem

Bacterial Resistance

Enterobacteriaceace

Enterobacteriaceace are common pathogens responsible for urinary tract infections, abdominal infections, and hospital-acquired pneumonia. Beta lactam resistance in these pathogens is most commonly due to the expression of beta lactamase enzymes.

Between 2007 and 2011, the percentage of *Escherichia coli* isolates from Canadian hospitals that produce extended spectrum beta lactamases (ESBL) increased from 3.4% to 4.1%; among *Klebsiella pneumoniae* isolates ESBL producers increased from 1.5% to 4.0%. These strains are resistant to third generation cephalosporins. that were developed for the treatment of beta lactamase-producing "Enterobacteriaceae" and carbapenems are generally regarded as the treatment of choice. More recently, many countries have experienced a dramatic upswing in the prevalence of *Enterobacteriaceace* that produce both ESBLs and carbapenemases such as the Klebsiella pneumoniae carbapenemase (KPC). As of 2013, 70% of Greek *Klebsiella pneumoniae* isolates are resistant to third generation and cyclosporins and 60% are resistant to carbapenems. The growing prevalence and difficulty of treating such multi-drug resistant *Enterobacteriaceace* has led to a renaissance of the use of antibiotics such as colistin, which was discovered in the 1950s but rarely used until recently due to unattractive levels of toxicity.

Pseudomonas Aeruginosa and Acinetobacter Baumannii

Infections caused by the non-fermenting gram-negative bacteria *Pseudomonas aeruginosa* and *Acinetobacter baumanni* are most commonly encountered in hospitalized people. These bacteria exhibit an unusually high level of intrinsic resistance to antibiotics due to their expression of a wide range of resistance mechanisms. Antibiotics cross the outer membrane of *Pseudomonas* and *Acinetobacter* approximately 100 times more slowly than they cross the outer membrane of *Enterobacteriaceae*, due in part to their use of porins that can adopt a conformation having a very restricted entry channel. Further, the porin levels may be down-regulated in response to antibiotic exposure. Antibiotic molecule that successfully traverse the porin channels may be removed by efflux pumps. Downregulation of the porin OprD2 is an important contributor to imipenem resistance.

Like the *Enterobacteriaceace*, *Pseudomonas* and *Acinetobacter* can express a wide range of antibiotic-deactivitating enzymes, including beta lactamases. *Pseudomonas* produces an inducible broad spectrum beta lactamase, AmpC, that is produced in response to beta lactam exposure. The combination of inducible AmpC expression, poor membrane permeability, and efflux pumps make *Pseudomonas* resistant to most beta lactams. The clinical efficacy of carbapenems in *Pseudomonas* infection arises in part because, while they are strong inducers of AmpC, they are poor substrates. The identification of *Pseudomonas* strains that produce beta lactamases capable of cleaving carbapenems, such as the New Delhi metallo beta lactamase has raised increasing concern regarding the potential for an era of untreatable *Pseudomonas* infections.

Structure

In terms of structure, the carbapenems are very similar to the penicillins (penams), but the sulfur atom in position 1 of the structure has been replaced with a carbon atom, and an unsaturation has been introduced—hence the name of the group, the carbapenems.

Biosynthesis

The carbapenems are thought to share their early biosynthetic steps in which the core ring system is formed. Malonyl-CoA is condensed with glutamate-5-semialdehyde with concurrent formation of the five-membered ring. Next, a β-lactam synthetase uses ATP to form the β-lactam and the saturated carbapenam core. Further oxidation and ring inversion provides the basic carbapenem.

Administration

Due to their expanded spectra, the desire to avoid generation of resistance and the fact that, in general, they have poor oral bioavailability, they are administered intravenously in hospital settings for more serious infections. However, research is underway to develop an effective oral carbapenem.

Tigecycline

Tigecycline (INN) is an antibiotic used to treat a number of bacterial infections. It is a first-in-class glycylcycline that is administered intravenously. It was developed in response to the growing rate of antibiotic resistance in bacteria such as *Staphylococcus aureus*, *Acinetobacter baumannii*, and *E. coli*. As a tetracycline derivative antibiotic, its structural modifications has expanded its therapeutic activity to include Gram-positive and Gram-negative organisms, including those of multidrug resistance. It is approved to treat complicated skin and soft tissue infections (cSSTI), complicated intra-abdominal infections (cIAI), and community-acquired bacterial pneumonia (CAP) in individuals 18 years and older.

Tigecycline is marketed by Pfizer under the brand name Tygacil. It was given a U.S. Food and Drug Administration (FDA) fast-track approval and was approved on June 17, 2005.

Medical Uses

Tigecycline is used to treat different kinds of bacterial infections, including complicated skin and structure infections, complicated intra-abdominal infections and community-acquired bacterial pneumonia. The spectrum of activity of tigecycline is discussed below.

Tigecycline can treat complicated skin and structure infections caused by; *Escherichia coli*, vancomycin-susceptible *Enterococcus faecalis*, methicillin-resistant *Staphylococcus aureus* (MRSA), *Streptococcus agalactiae*, *Streptococcus anginosus* grp., *Streptococcus pyogenes*, *Enterobacter cloacae*, *Klebsiella pneumoniae*, and *Bacteroides fragilis*.

Tigecycline is indicated for treatment of complicated intra-abdominal infections caused by; *Citrobacter freundii*, *Enterobacter cloacae*, *Escherichia coli*, *Klebsiella oxytoca*, *Klebsiella pneumoniae*, vancomycin-susceptible *Enterococcus faecalis*, methicillin-resistant *Staphylococcus aureus* (MRSA), *Streptococcus anginosus* grp., *Bacteroides fragilis*, *Bacteroides thetaiotaomicron*, *Bacteroides uniformis*, *Bacteroides vulgatus*, *Clostridium perfringens*, and *Peptostreptococcus micros*.

Tigecycline may be used for treatment of community-acquired bacterial pneumonia caused by; penicillin susceptible *Streptococcus pneumoniae*, *Haemophilus influenzae* that does not produce Beta-lactamase and *Legionella pneumophila*.

Tigecycline is given intravenously and has activity against a variety of Gram-positive and Gram-negative bacterial pathogens, many of which are resistant to existing antibiotics. Tigecycline successfully completed phase III trials in which it was at least equal to intravenous Vancomycin and Aztreonam to treat complicated skin and skin structure infections, and to intravenous imipenem and cilastatin to treat complicated intra-abdominal infections. Tigecycline is active against many Gram-positive bacteria, Gram-negative bacteria and anaerobes – including activity against methicillin-resistant *Staphylococcus aureus* (MRSA), *Stenotrophomonas maltophilia*, *Haemophilus influenzae*, and *Neisseria gonorrhoeae* (with MIC values reported at 2 µg/mL) and multi-drug resistant strains of *Acinetobacter baumannii*. It has no activity against *Pseudomonas* spp. or *Proteus* spp. The drug is licensed for the treatment of skin and soft tissue infections as well as intra-abdominal infections.

The European Society of Clinical Microbiology and Infection recommends tigecycline as a potential salvage therapy for severe and/or complicated or refractory Clostridium difficile infection.

Tigecycline can also be used in vulnerable populations such as immunocompromised patients or patients with cancer. Tigecycline may also have potential for use in Acute myeloid leukemia.

Susceptibility Data

Tigecycline targets both Gram-positive and Gram-negative bacteria including a few key multi-drug resistant pathogens. The following represents MIC susceptibility data for a few medically significant bacterial pathogens.

- *Escherichia coli*: 0.015 µg/ml - 4 µg/ml

- *Klebsiella pneumoniae*: 0.06 µg/ml - 16 µg/ml

- *Staphylococcus aureus* (methicillin-resistant): 0.03 µg/ml - 2 µg/ml

Tigecycline generally has poor activity against most strains of Pseudomonas.

Dosing

Tigecycline is given by slow intravenous infusion (30 to 60 minutes) every 12 hours. Patients with impaired liver function need to be given a lower dose. No adjustment is needed for patients with impaired kidney function. It is not licensed for use in children. There is no oral form available.

Liver or Kidney Problems

Tigecycline does not require dose adjustment for people with mild to moderate liver problems. However, in people with severe liver problems dosing should be decreased and closely monitored.

Tigecycline does not require dose changes in people with poor kidney function or having hemodialysis.

Side Effects

As a tetracycline derivative, tigecycline exhibits similar side effects to the class of antibiotics. Gastrointestinal (GI) symptoms are the most common reported side effect.

Common side effects of tigecycline include nausea and vomiting. Nausea (26%) and vomiting (18%) tend to be mild or moderate and usually occur during the first two days of therapy.

Rare adverse effects (<2%) include: swelling, pain, and irritation at injection site, anorexia, jaundice, hepatic dysfunction, pruritus, acute pancreatitis, and increased prothrombin time.

Precautions

Precaution is needed when taken in individuals with tetracycline hypersensitivity, pregnant women, and children. It has been found to cause fetal harm when administered during pregnancy and therefore is classified as pregnancy category D. In rats or rabbits, tigecycline crossed the placenta and was found in the fetal tissues, and is associated with slightly lower birth weights as well as slower bone ossification. Even though it was not considered teratogenic, tigecycline should be avoided unless benefits outweigh the risks. In addition, its use during childhood can cause yellow-grey-brown discoloration of the teeth and should not be used unless necessary.

More so, there are clinical reports of tigecycline-induced acute pancreatitis, with particular relevance to patients also diagnosed with cystic fibrosis.

Tigecycline showed an *increased* mortality in patients treated for hospital-acquired pneumonia, especially ventilator-associated pneumonia (a non-approved use), but also in patients with complicated skin and skin structure infections, complicated intra-abdominal infections and diabetic foot infection. Increased mortality was in comparison to other treatment of the same types of infections. The difference was not statistically significant for any type, but mortality was numerically greater for every infection type with Tigecycline treatment, and prompted a black box warning by the FDA.

Black Box Warning

FDA issued a black box warning in September 2010 for tigecycline regarding an increased risk of death compared to other appropriate treatment. As a result of increase in total death rate (cause is unknown) in individuals taking this drug, tigecycline is reserved for situations in which alternative treatment is not suitable.

In 2010, the U.S. Food and Drug Administration (FDA) updated the warnings section of the drug label to include information regarding increased mortality risk (seen most clearly in people treated for hospital-acquired pneumonia, especially ventilator-associated pneumonia).

Drug Interactions

Tigecycline has been found to interact with medications, such as:

- Warfarin: Since both tigecycline and warfarin bind to serum or plasma proteins, there is potential for protein-binding interactions, such that one drug will have more effect than the other. Although dose adjustment is not necessary, INR and prothrombin time should be monitored if given concurrently.

- Oral contraceptives: Effectiveness of oral contraceptives are decreased with concurrent use due to reduction in the concentration levels of oral contraceptives.

However, the mechanism behind these drug interactions have not been fully analyzed.

Mechanism of Action

Tigecycline is broad-spectrum antibiotic that acts as a protein synthesis inhibitor. It exhibits bacteriostatic activity by binding to the 30S ribosomal subunit of bacteria and thereby blocking the interaction of aminoacyl-tRNA with the A site of the ribosome. In addition, tigecycline has demonstrated bactericidal activity against isolates of *S. pneumoniae* and *L. pneumophila*.

It is a third generation tetracycline derivative within a class called glycylcyclines which carry a N,N-dimethyglycylamido (DMG) moiety attached to the 9-position of tetracycline ring D. With structural modifications as a 9-DMG derivative of minocycline, tigecycline has been found to improve minimal inhibitory concentrations against Gram-negative and Gram-positive organisms, when compared to tetracyclines.

Pharmacokinetics

Tigecycline is metabolized through glucuronidation into glucuronid conjugates and N-acetyl-9-aminominocycline metabolite. Therefore, dose adjustments are needed for patients with severe hepatic impairment. More so, it is primarily eliminated unchanged in the feces and secondarily eliminated by the kidneys. No renal adjustments are necessary.

Daptomycin

Daptomycin is a lipopeptide antibiotic used in the treatment of systemic and life-threatening infections caused by Gram-positive organisms. It is a naturally occurring compound found in the soil saprotroph *Streptomyces roseosporus*. Its distinct mechanism of action makes it useful in treating infections caused by multiple drug-resistant bacteria. It is marketed in the United States under the trade name Cubicin by Cubist Pharmaceuticals.

History

The compound LY 146032 was discovered by researchers at Eli Lilly and Company in the late 1980s. LY 146032 showed promise in phase I/II clinical trials for treatment of infection caused by Gram-positive organisms. Lilly ceased development because high-dose therapy was associated with adverse effects on skeletal muscle, including myalgia and potential myositis.

The rights to LY 146032 were acquired by Cubist Pharmaceuticals in 1997, which following U.S. Food and Drug Administration (FDA) approval in September 2003 for use in people older than 18 years, began marketing the drug under the trade name Cubicin. Cubicin is marketed in the EU and in several other countries by Novartis following its purchase of Chiron Corporation, the previous licensee.

Mechanism of Action

Daptomycin has a distinct mechanism of action, disrupting multiple aspects of bacterial cell membrane function. It inserts into the cell membrane in a phosphatidylglycerol-dependent fashion, where it then aggregates. The aggregation of daptomycin alters the curvature of the membrane, which creates holes that leak ions. This causes rapid depolarization, resulting in a loss of membrane potential leading to inhibition of protein, DNA, and RNA synthesis, which results in bacterial cell death.

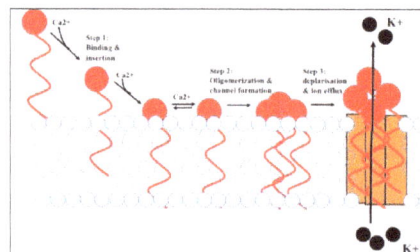

1. Daptomycin binds and inserts into the cell membrane. 2. Aggregates in the cell membrane. 3. Alters the shape of the cell membrane to form a hole in the cell, allowing ions in and out of the cell easily.

Microbiology

Daptomycin is bactericidal against Gram-positive bacteria only. It has proven *in vitro* activity against enterococci (including glycopeptide-resistant enterococci (GRE)), staphylococci (including methicillin-resistant *Staphylococcus aureus*), streptococci, corynebacteria and stationary-phase *Borrelia burgdorferi* persisters.

Daptomycin Resistance

Daptomycin resistance is still uncommon, but has been increasingly reported in GRE, starting in Korea in 2005, in Europe in 2010, in Taiwan 2011, and in the USA, where nine cases have been reported from 2007 to 2011. Daptomycin resistance emerged in five of the six cases while they were treated. The mechanism of resistance is unknown.

Clinical Use

Indications

Daptomycin is approved for use in adults in the United States for skin and skin structure infections caused by Gram-positive infections, *S. aureus* bacteraemia, and right-sided *S. aureus* endocarditis. It binds avidly to pulmonary surfactant, so cannot be used in the treatment of pneumonia. There seems to be a difference in working daptomycin on hematogenous pneumonia.

Efficacy

Daptomycin has been shown to be non-inferior to standard therapies (nafcillin, oxacillin, flucloxacillin or vancomycin) in the treatment of bacteraemia and right-sided endocarditis caused by *S. aureus*. A study in Detroit, Michigan compared 53 patients treated for suspected MRSA skin or soft tissue infection with daptomycin against vancomycin, showing faster recovery (4 versus 7 days) with daptomycin.

In phase III clinical trials, limited data showed daptomycin to be associated with poor outcomes in patients with left-sided endocarditis.. Daptomycin has not been studied in patients with prosthetic valve endocarditis or meningitis.

Dosage and Presentation

In skin and soft tissue infections, 4 mg/kg daptomycin is given intravenously once daily. For *S. aureus* bacteraemia or right-sided endocarditis, the approved dose is 6 mg/kg given intravenously once daily.

Daptomycin is given every 48 hours in patients with renal impairment with a creatinine clearance of less than 30 ml/min. No information is available on dosing in people less than 18 years of age.

Daptomycin is supplied as a sterile, preservative-free, pale yellow to light brown, lyophilised 500- or 350-mg cake that must be reconstituted with normal saline prior to use.

Daptomycin is applicable as 30-min infusion or 2-min injection.

Adverse Effects

Common adverse drug reactions associated with daptomycin therapy include:

- Cardiovascular: hypotension, hypertension, edema
- Central nervous system: insomnia

- Dermatological: rash

- Gastrointestinal: diarrhea, abdominal pain

- Hematological: eosinophilia

- Respiratory: dyspnea

- Other: injection site reactions, fever, hypersensitivity

Also, myopathy and rhabdomyolysis have been reported in patients simultaneously taking statins, but whether this is due entirely to the statin or whether daptomycin potentiates this effect is unknown. Due to the limited data available, the manufacturer recommends that statins be temporarily discontinued while the patient is receiving daptomycin therapy. Creatine kinase levels are usually checked regularly while individuals undergo daptomycin therapy.

In July 2010, the FDA issued a warning that daptomycin could cause life-threatening eosinophilic pneumonia. The FDA said it had identified seven confirmed cases of eosinophilic pneumonia between 2004 and 2010 and an additional 36 possible cases. The seven confirmed victims were all older than 60 and symptoms appeared within two weeks of initiation of therapy.

Biosynthesis

Figure 8. Structures of lipopeptide antibiotics Colors highlight the positions in daptomycin that have been modified by genetic engineering, as well as the origins of modules or subunits from A54145 or calcium-dependent antibiotic (CDA).

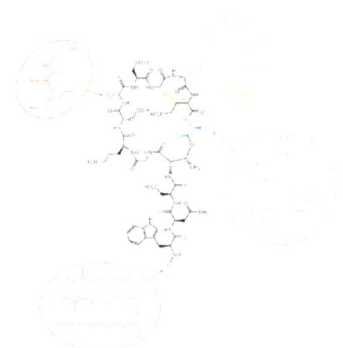

Figure 9. Combinatorial biosynthesis of lipopeptide antibiotics related to daptomycin. Position 8, which typically has D-Ala in daptomycin, was modified by module exchanges to contain D-Ser, D-Asn or D-Lys; position 11, which naturally has D-Ser, was modified by module exchanges to consist of D-Ala or D-Asn; position 12, which normally has 3-methyl-L-Glu, was modified by deletion of the methyltransferase gene to possess L-Glu; position 13, which normally has L-kynurenine (L-Kyn), was modified by subunit exchanges to contain L-Trp, L-Ile or L-Val; position 1 usually includes the anteiso-undecanoyl, isododecanoyl and anteiso-tridecanoyl fatty acyl groups. All of these alterations have been combinatorialized.

Daptomycin is a cyclic lipopeptide antibiotic produced by *Streptomyces roseosporus*. Daptomycin consists of 13 amino acids, 10 of which are arranged in a cyclic fashion, and three on an exocyclic tail. Two nonproteinogenic amino acids exist in the lipopeptide, the unusual amino acid L-kynurenine (Kyn), only known to daptomycin, and L-3-methylglutamic acid (mGlu). The N-terminus of the exocyclic tryptophan residue is coupled to decanoic acid, a medium-chain (C10) fatty acid. Biosynthesis is initiated by the coupling of decanoic acid to the N-terminal tryptophan, followed by the coupling of the remaining amino acids by nonribosomal peptide synthetase (NRPS) mechanisms. Finally, a cyclization event occurs, which is catalyzed by a thioesterase enzyme, and subsequent release of the lipopeptide is granted.

The NRPS responsible for the synthesis of daptomycin is encoded by three overlapping genes, *dptA, dptBC* and *dptD*. The *dptE* and *dptF* genes, immediately upstream of *dptA*, are likely to be involved in the initiation of daptomycin biosynthesis by coupling decanoic acid to the N-terminal Trp. These novel genes (dptE, dptF) correspond to products that most likely work in conjunction with a unique condensation domain to acylate the first amino acid (tryptophan). These and other novel genes (*dptI, dptJ*) are believed to be involved in supplying the nonproteinogenic amino acids L-3-methylglutamic acid and Kyn; they are located next to the NRPS genes.

The decanoic acid portion of daptomycin is synthesized by fatty acid synthase machinery. Post-translational modification of the apo-acyl carrier protein (ACP, thiolation, or T do-main) by a phosphopantetheinyltransferase (PPTase) enzyme catalyzes the transfer of a flexible phosphopantetheine arm from coenzyme A to a conserved serine in the ACP domain through a phosphodiester linkage. The holo-ACP can provide a thiol on which the substrate and acyl chains are covalently bound during chain elongations. The two core catalytic domains are an acyltransferase (AT) and a ketosynthase (KS). The AT acts upon a malonyl-CoA substrate and transfers an acyl group to the thiol of the ACP domain. This net transthiolation is an energy-neutral step. Next, the acyl-S-ACP gets transthiolated to a conserved cysteine on the KS; the KS decarboxylates the downstream malonyl-S-ACP and forms a β-ketoacyl-S-ACP. This serves as the substrate for the next cycle of elongation. Before the next cycle begins, however, the β-keto group undergoes reduction to the corresponding alcohol catalyzed by a ketoreductase domain, followed by dehydration to the olefin catalyzed by a dehydratase domain, and finally reduction to the methylene catalyzed by an enoylreductase domain. Each KS catalytic cycle results in the net addition of two carbons. After three more iterations of elongation, a thioesterase enzyme catalyzes the hydrolysis, and thus release, of the free C-10 fatty acid.

To synthesize the peptide portion of daptomycin, the mechanism of an NRPS is employed. The biosynthetic machinery of an NRPS system is composed of multimodular enzymatic assembly lines that contain one module for each amino acid monomer incorporated. Within each module are catalytic domains that carry out the elongation of the growing peptidyl chain. The growing peptide is covalently tethered to a thiolation domain; here it is termed the peptidyl carrier protein, as it carries the growing peptide from one catalytic domain to the next. Again, the apo-T domain must be primed to the holo-T domain by a PPTase, attaching a flexible phosphopantetheine arm to a conserved serine residue. An adenylation domain selects the amino acid monomer to be incorporated and activates the carboxylate with ATP to make the aminoacyl-AMP. Next, the A domain installs an aminoacyl group on the thiolate of the adjacent T domain. The condensation (C) domain catalyzes the peptide bond forming reaction, which elicits chain elongation. It joins an upstream peptidyl-S-T to the downstream aminoacyl-S-T. Chain elongation by one aminoacyl residue and chain translocation to the next T domain occurs in concert. The order of these domains is C-A-T. In some instances, an epimerization domain is necessary in those modules where L-amino acid monomers are to be incorporated and epimerized to D-amino acids. The domain organization in such modules is C-A-T-E.

The first module has a three-domain C-A-T organization; these often occur in assembly lines that make N-acylated peptides. The first C domain catalyzes N-acylation of the initiating amino acid (tryptophan) while it is installed on T. An adenylating enzyme (Ad) catalyzes the condensation of decanoic acid and the N-terminal tryptophan, which incorporates decanoic acid into the growing peptide. The genes responsible for this coupling event are dptE and dptF, which are located upstream of dptA, the first gene of the Daptomycin NRPS biosynthetic gene cluster. Once the coupling of decanoic acid to the N-terminal tryptophan residue occurs, the condensation of amino acids begins, catalyzed by the NRPS.

The first five modules of the NRPS are encoded by the *dptA* gene and catalyze the condensation of L-tryptophan, D-asparagine, L-aspartate, L-threonine, and glycine, respectively. Modules 6-11, which catalyze the condensation of L-ornithine, L-aspartate, D-alanine, L-aspartate, glycine, and D-serine are encoded for the *dptBC* gene. *dptD* catalyzes the incorporation of two nonproteinogenic amino acids, L-3-methylglutamic acid (mGlu) and Kyn, which is only known thus far to daptomycin, into the growing peptide. Elongation by these NRPS modules ultimately leads to macrocyclization and release in which an α-amino group, namely threonine, acts as an internal nucleophile during cyclization to yield the 10-amino-acid ring. The termination module in the NRPS assembly line has a C-A-T-TE organization. The thioesterase domain catalyzes chain termination and release of the mature lipopeptide.

The molecular engineering of daptomycin, the only marketed acidic lipopeptide antibiotic to date (Figure 8), has seen many advances since its inception into clinical medicine in 2003. It is an attractive target for combinatorial biosynthesis for many reasons: second generation derivatives are currently in the clinic for development; *Streptomyces roseosporus*, the producer organism of daptomycin, is amenable to genetic manipulation; the daptomycin biosynthetic gene cluster has been cloned, sequenced, and expressed in *S. lividans*; the lipopeptide biosynthetic machinery has the potential to be interrupted by variations of natural precursors, as well as precursor-directed biosynthesis, gene deletion, genetic exchange, and module exchange; the molecular engineering tools have been developed to facilitate the expression of the three individual NRPS genes from three different sites in the chromosome, using ermEp* for expression of two genes from ectopic loci; other lipopeptide gene clusters, both related and unrelated to daptomycin, have been cloned and sequenced, thus providing genes and modules to allow the generation of hybrid molecules; derivatives can be afforded via chemoenzymatic synthesis; and lastly, efforts in medicinal chemistry are able to further modify these products of molecular engineering.

New derivatives of daptomycin (Figure 9) were originally generated by exchanging the third NRPS subunit (*dptD*) with the terminal subunits from the A54145 (Factor B1) or calcium-dependent antibiotic pathways to create molecules containing Trp13, Ile13, or Val13. *dptD* is responsible for incorporating the penultimate amino acid, 3-methyl-glutamic acid (3mGlu12), and the last amino acid, Kyn13, into the chain. This exchange was achieved without engineering the interpeptide docking sites. These whole-subunit exchanges have been coupled with the deletion of the

Glu12-methyltransferase gene, with module exchanges at intradomain linker sites at Ala8 and Ser11, and with variations of natural fatty-acid side chains to generate over 70 novel lipopeptides in significant quantities; most of these resultant lipopeptides have potent antibacterial activities. Some of these compounds have *in vitro* antibacterial activities analogous to daptomycin. Further, one displayed ameliorated activity against an *E. coli* imp mutant that was defective in its ability to assemble its inherent lipopolysaccharide. A number of these compounds were produced in yields that spanned from 100 to 250 mg/liter; this, of course, opens up the possibility for successful scale-ups by fermentation techniques. Only a small percentage of the possible combinations of amino acids within the peptide core have been investigated thus far.

Vaccination

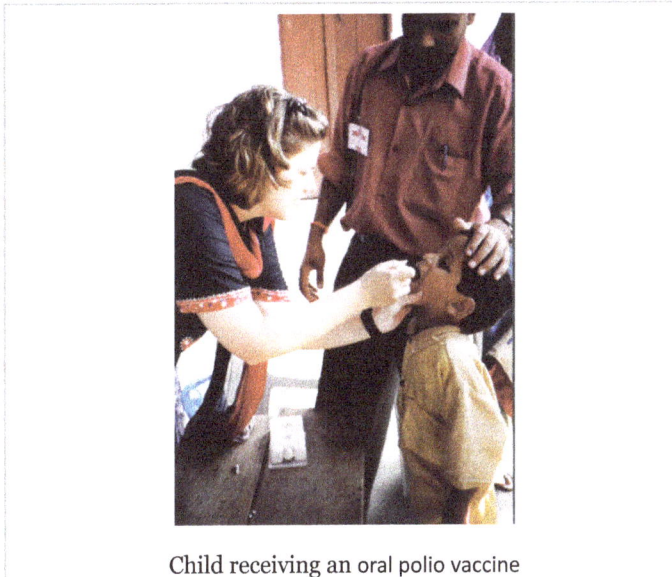

Child receiving an oral polio vaccine

Vaccination is the administration of antigenic material (a vaccine) to stimulate an individual's immune system to develop adaptive immunity to a pathogen. Vaccines can prevent or ameliorate morbidity from infection. When a sufficiently large percentage of a population has been vaccinated, this results in herd immunity. The effectiveness of vaccination has been widely studied and verified; for example, the influenza vaccine, the HPV vaccine, and the chicken pox vaccine. Vaccination is the most effective method of preventing infectious diseases; widespread immunity due to vaccination is largely responsible for the worldwide eradication of smallpox and the restriction of diseases such as polio, measles, and tetanus from much of the world. The World Health Organization (WHO) reports that licensed vaccines are currently available to prevent or contribute to the prevention and control of twenty-five preventable infections.

The active agent of a vaccine may be intact but inactivated (non-infective) or attenuated (with reduced infectivity) forms of the causative pathogens, or purified components of the pathogen that have been found to be highly immunogenic (e.g., outer coat proteins of a virus). Toxoids are produced for immunization against toxin-based diseases, such as the modification of tetanospasmin toxin of tetanus to remove its toxic effect but retain its immunogenic effect.

Smallpox was most likely the first disease people tried to prevent by inoculating themselves and was the first disease for which a vaccine was produced. The smallpox vaccine was discovered in 1796 by the British physician Edward Jenner and although at least six people had used the same principles years earlier he was the first to publish evidence that it was effective and to provide advice on its production. . Louis Pasteur furthered the concept through his work in microbiology. The immunization was called *vaccination* because it was derived from a virus affecting cows. Smallpox was a contagious and deadly disease, causing the deaths of 20–60% of infected adults and over 80% of infected children. When smallpox was finally eradicated in 1979, it had already killed an estimated 300–500 million people in the 20th century.

In common speech, *vaccination* and *immunization* have a similar meaning. This distinguishes it from inoculation, which uses unweakened live pathogens, although in common usage either can refer to an immunization. Vaccination efforts have been met with some controversy on scientific, ethical, political, medical safety, and religious grounds. In rare cases, vaccinations can injure people and, in the United States, they may receive compensation for those injuries under the National Vaccine Injury Compensation Program. Early success and compulsion brought widespread acceptance, and mass vaccination campaigns have greatly reduced the incidence of many diseases in numerous geographic regions.

Mechanism of Function

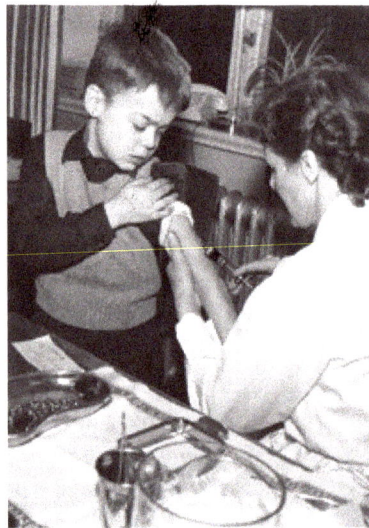

Polio vaccination started in Sweden in 1957.

Generically, the process of artificial induction of immunity, in an effort to protect against infectious disease, works by 'priming' the immune system with an 'immunogen'. Stimulating immune responses with an infectious agent is known as *immunization*. Vaccination includes various ways of administering immunogens.

Some vaccines are administered after the patient already has contracted a disease. Vaccines given after exposure to smallpox, within the first three days, are reported to attenuate the disease considerably, and vaccination up to a week after exposure probably offers some protection from disease or may modify the severity of disease. The first rabies immunization was given by Louis Pasteur to a child after he was bitten by a rabid dog. Subsequent to this, it has been found that, in people with uncompromised immune systems, four doses of rabies vaccine over 14 days, wound care, and treatment of the bite with rabies immune globulin, commenced as soon as possible after exposure, is effective in preventing the development of rabies in humans. Other examples include experimental AIDS, cancer and Alzheimer's disease vaccines. Such immunizations aim to trigger an immune response more rapidly and with less harm than natural infection.

Most vaccines are given by hypodermic injection as they are not absorbed reliably through the intestines. Live attenuated polio, some typhoid, and some cholera vaccines are given orally to produce immunity in the bowel. While vaccination provides a lasting effect, it usually takes several weeks to develop, while passive immunity (the transfer of antibodies) has immediate effect.

Adjuvants and Preservatives

Vaccines typically contain one or more adjuvants, used to boost the immune response. Tetanus toxoid, for instance, is usually adsorbed onto alum. This presents the antigen in such a way as to produce a greater action than the simple aqueous tetanus toxoid. People who get an excessive reaction to adsorbed tetanus toxoid may be given the simple vaccine when time for a booster occurs.

In the preparation for the 1990 Persian Gulf campaign, pertussis vaccine (not acellular) was used as an adjuvant for anthrax vaccine. This produces a more rapid immune response than giving only the anthrax, which is of some benefit if exposure might be imminent.

Vaccines may also contain preservatives to prevent contamination with bacteria or fungi. Until recent years, the preservative thimerosal was used in many vaccines that did not contain live virus. As of 2005, the only childhood vaccine in the U.S. that contains thimerosal in greater than trace amounts is the influenza vaccine, which is currently recommended only for children with certain risk factors. Single-dose influenza vaccines supplied in the UK do not list thiomersal (its UK name) in the ingredients. Preservatives may be used at various stages of production of vaccines, and the most sophisticated methods of measurement might detect traces of them in the finished product, as they may in the environment and population as a whole.

Vaccination Versus Inoculation

Many times these words are used interchangeably, as if they were synonyms. In fact, they are different things. As doctor Byron Plant explains: "Vaccination is the more commonly used term, which actually consists of a 'safe' injection of a sample taken from a cow suffering from cowpox... Inoculation, a practice probably as old as the disease itself, is the injection of the variola virus taken from a pustule or scab of a smallpox sufferer into the superficial layers of the skin, commonly on the upper arm of the subject. Often inoculation was done 'arm to arm' or less effectively 'scab to arm'..." Inoculation oftentimes caused the patient to become infected with smallpox, and in some cases the infection turned into a severe case.

Vaccination began in the 18th century with the work of Edward Jenner and the smallpox vaccine.

Types

Vaccines work by presenting a foreign antigen to the immune system to evoke an immune response, but there are several ways to do this. Four main types are currently in clinical use:

1. An inactivated vaccine consists of virus or bacteria that are grown in culture and then killed using a method such as heat or formaldehyde. Although the virus or bacteria particles are destroyed and cannot replicate, the virus capsid proteins or bacterial wall are intact enough to be recognized and remembered by the immune system and evoke a response. When manufactured correctly, the vaccine is not infectious, but improper inactivation can result in intact and infectious particles. Since the properly produced vaccine does not reproduce, booster shots are required periodically to reinforce the immune response.

2. In an attenuated vaccine, live virus or bacteria with very low virulence are administered. They will replicate, but locally or very slowly. Since they do reproduce and continue to present antigen to the immune system beyond the initial vaccination, boosters may be required less often. These vaccines may be produced by passaging, for example, adapting a virus into different host cell cultures, such as in animals, or at suboptimal temperatures, allowing selection of less virulent strains, or by mutagenesis or targeted deletions in genes required for virulence. There is a small risk of reversion to virulence, which is smaller in vaccines with deletions. Attenuated vaccines also cannot be used by immunocompromised individuals. Reversions of virulence were described for a few attenuated viruses of chickens (infectious bursal disease virus, avian infectious bronchitis virus, avian infectious laryngotracheitis virus , avian metapneumovirus)

3. Virus-like particle vaccines consist of viral protein(s) derived from the structural proteins of a virus. These proteins can self-assemble into particles that resemble the virus from which they were derived but lack viral nucleic acid, meaning that they are not infectious. Because of their highly repetitive, multivalent structure, virus-like particles are typically more immunogenic than subunit vaccines (described below). The human papillomavirus and Hepatitis B virus vaccines are two virus-like particle-based vaccines currently in clinical use.

4. A subunit vaccine presents an antigen to the immune system without introducing viral particles, whole or otherwise. One method of production involves isolation of a specific protein from a virus or bacterium (such as a bacterial toxin) and administering this by itself. A weakness of this technique is that isolated proteins may have a different three-dimensional structure than the protein in its normal context, and will induce antibodies that may not recognize the infectious organism. In addition, subunit vaccines often elicit weaker antibody responses than the other classes of vaccines.

A number of other vaccine strategies are under experimental investigation. These include DNA vaccination and recombinant viral vectors.

History

It is known that the process of inoculation was used by Chinese physicians in the 10th century. Scholar Ole Lund comments: "The earliest documented examples of vaccination are from India and China in the 17th century, where vaccination with powdered scabs from people infected with smallpox was used to protect against the disease. Smallpox used to be a common disease throughout the world and 20 to 30% of infected persons died from the disease. Smallpox was responsible for 8 to 20% of all deaths in several European countries in the 18th century. The tradition of vaccination may have originated in India in AD 1000." The mention of inoculation in the *Sact'eya Grantham*, an Ayurvedic text, was noted by the French scholar Henri Marie Husson in the journal

Dictionaire des sciences médicales. Inoculation was reportedly widely practised in China in the reign of the Longqing Emperor (r. 1567–1572) during the Ming Dynasty (1368–1644). The Anatolian Ottoman Turks knew about methods of inoculation. This kind of inoculation and other forms of variolation were introduced into England by Lady Montagu, a famous English letter-writer and wife of the English ambassador at Istanbul between 1716 and 1718, who almost died from smallpox as a young adult and was physically scarred from it. Inoculation was adopted both in England and in America nearly half a century before Jenner's famous smallpox vaccine of 1796 but the death rate of about 2% from this method meant that it was mainly used during dangerous outbreaks of the disease and remained controversial.

Jenner's handwritten draft of the first vaccination

It was noticed during the 18th century that people who had suffered from the less virulent cowpox were immune to smallpox and the first recorded use of this idea was by a farmer Benjamin Jesty at Yetminster who had suffered the disease and transmitted it to his own family in 1774, his sons subsequently not getting the mild version of smallpox when later inoculated in 1789. But it was Edward Jenner, a doctor in Berkeley, who established the procedure by introducing material from a cowpox vesicle on Sarah Nelmes, a milkmaid, into the arm of a boy named James Phipps. Two months later he inoculated the boy with smallpox and the disease did not develop. In 1798, Jenner published "An Inquiry into the Causes and Effects of the Variolae Vacciniae" which coined the term *vaccination* and created widespread interest. He distinguished 'true' and 'spurious' cowpox (which did not give the desired effect) and developed an "arm-to-arm" method of propagating the vaccine from the vaccinated individual's pustule. Early attempts at confirmation were confounded by contamination with smallpox, but despite controversy within the medical profession and religious opposition to the use of animal material, by 1801 his report was translated into six languages and over 100,000 people were vaccinated.

Since then vaccination campaigns have spread throughout the globe, sometimes prescribed by law or regulations. Vaccines are now used against a wide variety of diseases besides smallpox. Louis Pasteur further developed the technique during the 19th century, extending its use to killed agents protecting against anthrax and rabies. The method Pasteur used entailed treating the agents for those diseases so they lost the ability to infect, whereas inoculation was the hopeful selection of a less virulent form of the disease, and Jenner's vaccination entailed the substitution of a different and less dangerous disease for the one protected against. Pasteur adopted the name *vaccine* as a generic term in honor of Jenner's discovery.

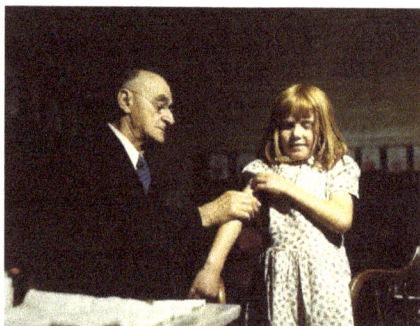

A doctor performing a typhoid vaccination in Texas, 1943

Maurice Hilleman was the most prolific vaccine inventor, and developed successful vaccines for measles, mumps, hepatitis A, hepatitis B, chickenpox, meningitis, pneumonia and *Haemophilus influenzae*.

In modern times, the first vaccine-preventable disease targeted for eradication was smallpox. The World Health Organization (WHO) coordinated this global eradication effort. The last naturally occurring case of smallpox occurred in Somalia in 1977. In 1988, the governing body of WHO targeted polio for eradication by 2000. Although the target was missed, eradication is very close. The next disease to be targeted for eradication would most likely be measles, which has declined since the introduction of measles vaccination in 1963.

In 2000, the Global Alliance for Vaccines and Immunization was established to strengthen routine vaccinations and introduce new and under-used vaccines in countries with a per capita GDP of under US$1000. GAVI is now entering its second phase of funding, which extends through 2014.

Society and Culture

Poster for vaccination against smallpox

To eliminate the risk of outbreaks of some diseases, at various times several governments and other institutions have employed policies requiring vaccination for all people. For example, an 1853 law required universal vaccination against smallpox in England and Wales, with fines levied on people who did not comply. Common contemporary U.S. vaccination policies require that children receive common vaccinations before entering public school.

Beginning with early vaccination in the nineteenth century, these policies were resisted by a variety of groups, collectively called antivaccinationists, who object on scientific, ethical, political, medical safety, religious, and other grounds. Common objections are that vaccinations do not work, that compulsory vaccination represents excessive government intervention in personal matters, or that the proposed vaccinations are not sufficiently safe. Many modern vaccination policies allow exemptions for people who have compromised immune systems, allergies to the components used in vaccinations or strongly held objections.

In countries with limited financial resources, limited coverage causes much morbidity and mortality. More affluent countries are able to subsidize vaccinations for at-risk groups, resulting in more comprehensive and effective coverage. In Australia, for example, the Government subsidizes vaccinations for seniors and indigenous Australians.

Public Health Law Research, an independent US based organization, reported in 2009 that there is insufficient evidence to assess the effectiveness of requiring vaccinations as a condition for specified jobs as a means of reducing incidence of specific diseases among particularly vulnerable populations; that there is sufficient evidence supporting the effectiveness of requiring vaccinations as a condition for attending child care facilities and schools; and that there is strong evidence supporting the effectiveness of standing orders, which allow healthcare workers without prescription authority to administer vaccine as a public health intervention aimed at increasing vaccination rates.

Vaccination-autism Controversy

In the MMR vaccine controversy, a fraudulent 1998 paper by Andrew Wakefield, originally published in *The Lancet*, presented supposed evidence that the MMR vaccine (an immunization against measles, mumps and rubella that is typically first administered to children shortly after their first birthday) was linked to the onset of autism spectrum disorders. The article was widely criticized for lack of scientific rigour, partially retracted in 2004 by Wakefield's co-authors, and was fully retracted by *The Lancet* in 2010. Wakefield was struck off the UK's medical registry for the fraud.

This Lancet article has sparked a much greater anti-vaccination movement, primarily in the United States. Even though the article was fraudulent and was retracted, 1 in 4 parents still believe vaccines can cause autism. Many parents do not vaccinate their children because they feel that diseases are no longer present due to all the vaccinations. This is a false assumption, since some diseases could still return. These pathogens could possibly infect vaccinated people, due to the pathogen's ability to mutate when it is able to live in unvaccinated hosts. In 2010, there was a whooping cough outbreak in California that was the worst outbreak in 50 years. A possible contributing factor was parents choosing to exempt their children from vaccinations. There was also a case in Texas in 2012 where 21 members of a church contracted measles because they chose to abstain from immunizations.

Side Effects and Injury

The Centers for Disease Control and Prevention (CDC) has compiled a list of vaccines and their possible side effects. Allegations of vaccine injuries in recent decades have appeared in litigation in the U.S. Some families have won substantial awards from sympathetic juries, even though most public health officials have said that the claims of injuries were unfounded. In response, several vaccine makers stopped production, which the US government believed could be a threat to public health, so laws were passed to shield makers from liabilities stemming from vaccine injury claims.

Routes of Administration

Air France Vaccinations Centre in the 7th arrondissement of Paris

A vaccine administration may be oral, by injection (intramuscular, intradermal, subcutaneous), by puncture, transdermal or intranasal. Several recent clinical trials have aimed to deliver the vaccines via mucosal surfaces to be up-taken by the common mucosal immunity system, thus avoiding the need for injections.

Global Trends in Vaccination

The World Health Organization (WHO) estimate that vaccination averts 2-3 million deaths per year (in all age groups), and up to 1.5 million children die each year due to diseases which could have been prevented by vaccination. They estimate that 29% of deaths of children under five years old in 2013 were vaccine preventable.

Vaccination in art

- *La vaccine* or *Le préjugé vaincu* by Louis-Léopold Boilly, 1807

- *A doctor vaccinating a small girl, other girls with loosened blouses wait their turn apprehensively* by Lance Calkin

- German caricature showing von Behring extracting the serum with a tap.

- *Les Malheurs de la Vaccine* (The history of vaccination seen from an economic point of view: A pharmacy up for sale; an outmoded inoculist selling his premises; Jenner, to the left, pursues a skeleton with a lancet)

Phage Therapy

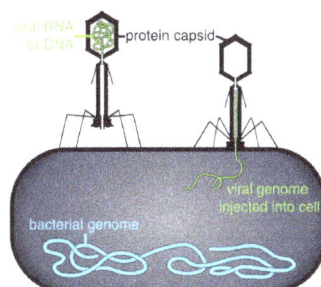

Phage injecting its genome into bacterial cell

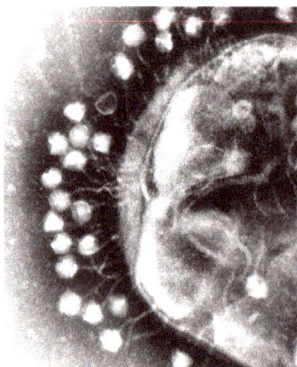

An electron micrograph of bacteriophages attached to a bacterial cell.
These viruses are the size and shape of coliphage T1.

Phage therapy or viral phage therapy is the therapeutic use of bacteriophages to treat pathogenic bacterial infections. Phage therapy has many potential applications in human medicine as well as dentistry, veterinary science, and agriculture. If the target host of a phage therapy treatment is not an animal, the term "biocontrol" (as in phage-mediated biocontrol of bacteria) is usually employed, rather than "phage therapy".

Bacteriophages are much more specific than antibiotics. They are typically harmless not only to the host organism, but also to other beneficial bacteria, such as the gut flora, reducing the chances of opportunistic infections. They have a high therapeutic index, that is, phage therapy would be expected to give rise to few side effects. Because phages replicate *in vivo*, a smaller effective dose can be used. On the other hand, this specificity is also a disadvantage: a phage will only kill a bacterium if it is a match to the specific strain. Consequently phage mixtures are often applied to improve the chances of success, or samples can be taken and an appropriate phage identified and grown.

Phages tend to be more successful than antibiotics where there is a biofilm covered by a polysaccharide layer, which antibiotics typically cannot penetrate. In the West, no therapies are currently authorized for use on humans, although phages for killing food poisoning bacteria (*Listeria*) are now in use.

Phages are currently being used therapeutically to treat bacterial infections that do not respond to conventional antibiotics, particularly in Russia and Georgia. There is also a phage therapy unit in Wroclaw, Poland, established 2005, the only such centre in a European Union country.

History

The discovery of bacteriophages was reported by Frederick Twort in 1915 and Felix d'Hérelle in 1917. D'Hérelle said that the phages always appeared in the stools of *Shigella* dysentery patients shortly before they began to recover. He "quickly learned that bacteriophages are found wherever bacteria thrive: in sewers, in rivers that catch waste runoff from pipes, and in the stools of convalescent patients." Phage therapy was immediately recognized by many to be a key way forward for the eradication of bacterial infections. A Georgian, George Eliava, was making similar discoveries. He travelled to the Pasteur Institute in Paris where he met d'Hérelle, and in 1923 he founded the Eliava Institute in Tbilisi, Georgia, devoted to the development of phage therapy. Phage therapy is used in Russia, Georgia and Poland.

Frederick Twort

Phage in action on Bacillus anthracis

In Russia, extensive research and development soon began in this field. In the United States during the 1940s commercialization of phage therapy was undertaken by Eli Lilly and Company.

While knowledge was being accumulated regarding the biology of phages and how to use phage cocktails correctly, early uses of phage therapy were often unreliable. When antibiotics were discovered in 1941 and marketed widely in the U.S. and Europe, Western scientists mostly lost interest in further use and study of phage therapy for some time.

Isolated from Western advances in antibiotic production in the 1940s, Russian scientists continued to develop already successful phage therapy to treat the wounds of soldiers in field hospitals. During World War II, the Soviet Union used bacteriophages to treat many soldiers infected with various bacterial diseases e.g. dysentery and gangrene. Russian researchers continued to develop and to refine their treatments and to publish their research and results. However, due to the scientific barriers of the Cold War, this knowledge was not translated and did not proliferate across the world. A summary of these publications was published in English in 2009 in "A Literature Review of the Practical Application of Bacteriophage Research".

There is an extensive library and research center at the George Eliava Institute in Tbilisi, Georgia. Phage therapy is today a widespread form of treatment in that region.

As a result of the development of antibiotic resistance since the 1950s and an advancement of scientific knowledge, there has been renewed interest worldwide in the ability of phage therapy to eradicate bacterial infections and chronic polymicrobial biofilm (including in industrial situations).

Phages have been investigated as a potential means to eliminate pathogens like *Campylobacter* in raw food and *Listeria* in fresh food or to reduce food spoilage bacteria. In agricultural practice phages were used to fight pathogens like *Campylobacter*, *Escherichia* and *Salmonella* in farm animals, *Lactococcus* and *Vibrio* pathogens in fish from aquaculture and *Erwinia* and *Xanthomonas* in plants of agricultural importance. The oldest use was, however, in human medicine. Phages have been used against diarrheal diseases caused by *E. coli*, *Shigella* or *Vibrio* and against wound infections caused by facultative pathogens of the skin like staphylococci and streptococci. Recently the phage therapy approach has been applied to systemic and even intracellular infections and the addition of non-replicating phage and isolated phage enzymes like lysins to the antimicrobial arsenal. However, actual proof for the efficacy of these phage approaches in the field or the hospital is not available.

Some of the interest in the West can be traced back to 1994, when Soothill demonstrated (in an animal model) that the use of phages could improve the success of skin grafts by reducing the underlying *Pseudomonas aeruginosa* infection. Recent studies have provided additional support for these findings in the model system.

Although not "phage therapy" in the original sense, the use of phages as delivery mechanisms for traditional antibiotics constitutes another possible therapeutic use. The use of phages to deliver antitumor agents has also been described in preliminary *in vitro* experiments for cells in tissue culture.

In June 2015 the European Medicines Agency hosted a one-day workshop on the therapeutic use of bacteriophages and in July 2015 the National Institutes of Health (USA) hosted a two-day workshop "Bacteriophage Therapy: An Alternative Strategy to Combat Drug Resistance".

Potential Benefits

Bacteriophage treatment offers a possible alternative to conventional antibiotic treatments for bacterial infection. It is conceivable that, although bacteria can develop resistance to phage, the resistance might be easier to overcome than resistance to antibiotics. Just as bacteria can evolve resistance, viruses can evolve to overcome resistance.

Bacteriophages are very specific, targeting only one or a few strains of bacteria. Traditional antibiotics have more wide-ranging effect, killing both harmful bacteria and useful bacteria such as those facilitating food digestion. The species and strain specificity of bacteriophages makes it unlikely that harmless or useful bacteria will be killed when fighting an infection.

Some evidence shows the ability of phages to travel to a required site—including the brain, where the blood brain barrier can be crossed—and multiply in the presence of an appropriate bacterial host, to combat infections such as meningitis. However the patient's immune system can, in some cases, mount an immune response to the phage (2 out of 44 patients in a Polish trial).

A few research groups in the West are engineering a broader spectrum phage, and also a variety of forms of MRSA treatments, including impregnated wound dressings, preventative treatment for burn victims, phage-impregnated sutures. Enzybiotics are a new development at Rockefeller University that create enzymes from phage. These show potential for preventing secondary bacterial infections, e.g. pneumonia developing in patients suffering from flu and otitis. Purified recombinant phage enzymes can be used as separate antibacterial agents in their own right.

For some bacteria, such as multiple-resistant Klebsiella pneumoniae, there are no effective non-toxic antibiotics, but killing of this bacteria via intraperitoneal, intravenous, or intranasal route of phages in vivo has been shown to work in laboratory tests.

Application

Collection

The simplest method of phage treatment involves collecting local samples of water likely to contain high quantities of bacteria and bacteriophages, for example effluent outlets, sewage and other sources. They can also be extracted from corpses. The samples are taken and applied to the bacteria that are to be destroyed which have been cultured on growth medium.

If the bacteria die, as usually happens, the mixture is centrifuged; the phages collect on the top of the mixture and can be drawn off.

The phage solutions are then tested to see which ones show growth suppression effects (lysogeny) or destruction (lysis) of the target bacteria. The phage showing lysis are then amplified on cultures of the target bacteria, passed through a filter to remove all but the phages, then distributed.

Treatment

Phages are "bacterium-specific" and it is therefore necessary in many cases to take a swab from the patient and culture it prior to treatment. Occasionally, isolation of therapeutic phages can require a few months to complete, but clinics generally keep supplies of phage cocktails for the most common bacterial strains in a geographical area.

Phages in practice are applied orally, topically on infected wounds or spread onto surfaces, or used during surgical procedures. Injection is rarely used, avoiding any risks of trace chemical contaminants that may be present from the bacteria amplification stage, and recognizing that the immune system naturally fights against viruses introduced into the bloodstream or lymphatic system.

The direct human use of phages is likely to be very safe; in August and October 2006, the United States Food and Drug Administration approved spraying meat and cheese with phages. The approval was for ListShield and Listex (phage preparations targeting *Listeria monocytogenes*). These were the first approvals granted by the FDA and USDA for phage-based applications.

Phage therapy has been attempted for the treatment of a variety of bacterial infections including: laryngitis, skin infections, dysentery, conjunctivitis, periodontitis, gingivitis, sinusitis, urinary tract infections and intestinal infections, burns, boils, poly-microbial biofilms on chronic wounds, ulcers and infected surgical sites.

In 2007 a Phase 1/2 clinical trial was completed at the Royal National Throat, Nose and Ear Hospital, London for *Pseudomonas aeruginosa* infections (otitis). Documentation of the Phase-1/Phase-2 study was published in August 2009 in the journal Clinical Otolaryngology.

Phase 1 clinical trials have now been completed in the Southwest Regional Wound Care Center, Lubbock, Texas for an approved cocktail of phages against bacteria, including *P. aeruginosa*, *Staphylococcus aureus* and *Escherichia coli* (better known as E. coli). The cocktail of phages for the clinical trials was developed and supplied by Intralytix.

Reviews of phage therapy indicate that more clinical and microbiological research is needed to meet current standards.

Administration

Phages can usually be freeze-dried and turned into pills without materially impacting efficiency. Temperature stability up to 55 °C and shelf lives of 14 months have been shown for some types of phages in pill form.

Application in liquid form is possible, stored preferably in refrigerated vials.

Oral administration works better when an antacid is included, as this increases the number of phages surviving passage through the stomach.

Topical administration often involves application to gauzes that are laid on the area to be treated.

Obstacles

The high bacterial strain specificity of phage therapy may make it necessary for clinics to make different cocktails for treatment of the same infection or disease because the bacterial components of such diseases may differ from region to region or even person to person.

In addition, due to the specificity of individual phages, for a high chance of success, a mixture of phages is often applied. This means that 'banks' containing many different phages must be kept and regularly updated with new phages.

Further, bacteria can evolve different receptors either before or during treatment; this can prevent the phages from completely eradicating the bacteria.

The need for banks of phages makes regulatory testing for safety harder and more expensive under current rules in most countries. Such a process would make it difficult for large-scale production of phage therapy. Additionally, patent issues (specifically on living organisms) may complicate distribution for pharmaceutical companies wishing to have exclusive rights over their "invention", which would discourage for-profit corporation from investing capital in the dissemination of this technology.

As has been known for at least thirty years, mycobacteria such as *Mycobacterium tuberculosis* have specific bacteriophages. No lytic phage has yet been discovered for *Clostridium difficile*, which is responsible for many nosocomial diseases, but some *temperate phages* (integrated in the genome, also called lysogenic) are known for this species; this opens encouraging avenues but with additional risks as discussed below.

To work the virus has to reach the site of the bacteria, and viruses can sometimes reach places antibiotics cannot. For example, jazz bassist Alfred Gertler got a bacterial infection in his bones after breaking an ankle. A physician in the U.S. told him that the foot must be amputated. He refused and was largely bed ridden for four years until phage therapy at the Eliava Institute in Tbilisi, Georgia, eliminated the bacterial infection. Then he had surgery to repair his ankle and resumed his career and family life.

Funding for phage therapy research and clinical trials is generally insufficient and difficult to obtain, since it is a lengthy and complex process to patent bacteriophage products. Scientists comment that 'the biggest hurdle is regulatory', whereas an official view is that individual phages would need proof individually because it would be too complicated to do as a combination, with many variables. Due to the specificity of phages, phage therapy would be most effective with a cocktail injection, which is generally rejected by the U.S. Food and Drug Administration (FDA). Researchers and observers predict that for phage therapy to be successful the FDA must change its regulatory stance on combination drug cocktails. Public awareness and education about phage therapy are generally limited to scientific or independent research rather than mainstream media.

The negative public perception of viruses may also play a role in the reluctance to embrace phage therapy.

Legislation

Approval of phage therapy for use in humans has not been given in Western countries with a few exceptions. Washington and Oregon law allows naturopathic physicians to use any therapy that is legal any place in the world on an experimental basis.

In Texas phages are considered natural substances and can be used in addition to (but not as a replacement for) traditional therapy; they've been used routinely in a wound care clinic in Lubbock, TX, since 2006.

In 2013 "the 20th biennial Evergreen International Phage Meeting ... conference drew 170 participants from 35 countries, including leaders of companies and institutes involved with human phage therapies from France, Australia, Georgia, Poland and the United States."

Safety

Much of the difficulty in obtaining regulatory approval is proving safety for using a self-replicating entity which has the capability to evolve.

As with antibiotic therapy and other methods of countering bacterial infections, endotoxins are released by the bacteria as they are destroyed within the patient (Herxheimer reaction). This can cause symptoms of fever; in extreme cases toxic shock (a problem also seen with antibiotics) is possible. Janakiraman Ramachandran argues that this complication can be avoided in those types of infection where this reaction is likely to occur by using genetically engineered bacteriophages which have had their gene responsible for producing endolysin removed. Without this gene the host bacterium still dies but remains intact because the lysis is disabled. On the other hand, this modification stops the exponential growth of phages, so one administered phage means one dead bacterial cell. Eventually these dead cells are consumed by the normal house-cleaning duties of the phagocytes, which utilise enzymes to break down the whole bacterium and its contents into harmless proteins, polysaccharides and lipids.

Temperate (or Lysogenic) bacteriophages are not generally used therapeutically, as this group can act as a way for bacteria to exchange DNA; this can help spread antibiotic resistance or even, theoretically, make the bacteria pathogenic. Carl Merril claimed that harmless strains of corynebacterium may have been converted into c. diphtheriae that "probably killed a third of all Europeans who came to North America in the seventeenth century". Fortunately, many phages seem to be lytic only with negligible probability of becoming lysogenic.

Efficacy

In Russia, mixed phage preparations may have a therapeutic efficacy of 50%. This equates to the complete cure of 50 of 100 patients with terminal antibiotic-resistant infection. The rate of only 50% is likely to be due to individual choices in admixtures and ineffective diagnosis of the causative agent of infection.

Other Animals

Brigham Young University is currently researching the use of phage therapy to treat American foulbrood in honeybees.

Cultural Impact

The 1925 novel and 1926 Pulitzer prize winner *Arrowsmith* used phage therapy as a plot point.

Greg Bear's 2002 novel *Vitals* features phage therapy, based on Soviet research, used to transfer genetic material.

The 2012 collection of military history essays about the changing role of women in warfare, "Women in War - from home front to front line" includes a chapter featuring phage therapy: "Chapter 17: Women who thawed the Cold War".

Antibiotic Misuse

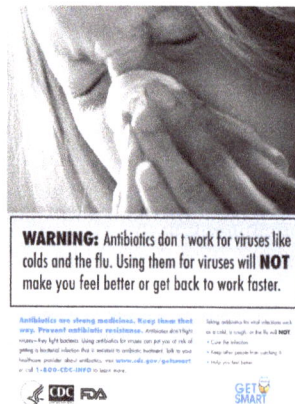

WARNING: Antibiotics don't work for viruses like colds and the flu. Using them for viruses will **NOT** make you feel better or get back to work faster.

This poster from the U.S. Centers for Disease Control and Prevention "Get Smart" campaign, intended for use in doctors' offices and other healthcare facilities, warns that antibiotics do not work for viral illnesses such as the common cold.

Antibiotic misuse, sometimes called antibiotic abuse or antibiotic overuse, refers to the misuse or overuse of antibiotics, with potentially serious effects on health. It is a contributing factor to the development of antibiotic resistance, including the creation of multidrug-resistant bacteria, informally called "super bugs": relatively harmless bacteria can develop resistance to multiple antibiotics and cause life-threatening infections.

Instances of Antibiotic Misuse

Health advocacy messages such as this one encourage patients to talk with their doctor about safety in using antibiotics.

Common situations in which antibiotics are overused include the following:

- Apparent viral respiratory illness in children should not be treated with antibiotics. If there is a diagnosis of bacterial infection, then antibiotics may be used.

- When children with ear tubes get ear infections, they should have antibiotic eardrops put into their ears to go to the infection rather than having oral antibiotics which are more likely to have unwanted side effects.

- Swimmer's ear should be treated with antibiotic eardrops, not oral antibiotics.

- Sinusitis should not be treated with antibiotics because it is usually caused by a virus, and even when it is caused by a bacteria, antibiotics are not indicated except in atypical circumstances as it usually resolves without treatment.

- Viral conjunctivitis should not be treated with antibiotics. Antibiotics should only be used with confirmation that a patient has bacterial conjunctivitis.

- Older persons often have bacteria in their urine which is detected in routine urine tests, but unless the person has the symptoms of a urinary tract infection, antibiotics should not be used in response.

- Eczema should not be treated with antibiotics. Dry skin can be treated with lotions or other symptom treatments.

- The use of antibiotics to treat surgical wounds does not reduce infection rates in comparison with non-antibiotic ointment or no ointment at all.

Social and Economic Impact

Antibiotics can cause severe reactions and add significantly to the cost of care. In the United States, antibiotics and anti-infectives are the leading cause of adverse effect from drugs. In a study of 32 States in 2011, antibiotics and anti-infectives accounted for nearly 24 percent of ADEs that were present on admission, and 28 percent of those that occurred during a hospital stay.

Prescribing by an infectious disease specialist compared with prescribing by a non-infectious disease specialist decreases antibiotic consumption and reduces costs.

Antibiotic Resistance

Though antibiotics are required to treat severe bacterial infections, misuse has contributed to a rise in bacterial resistance. The overuse of fluoroquinolone and other antibiotics fuels antibiotic resistance in bacteria, which can inhibit the treatment antibiotic-resistant infections. Their excessive use in children with otitis media has given rise to a breed of bacteria resistant to antibiotics entirely.

Widespread use of fluoroquinolones as a first-line antibiotic has led to decreased antibiotic sensitivity, with negative implications for serious bacterial infections such as those associated with cystic fibrosis, where quinolones are among the few viable antibiotics.

Inappropriate Use

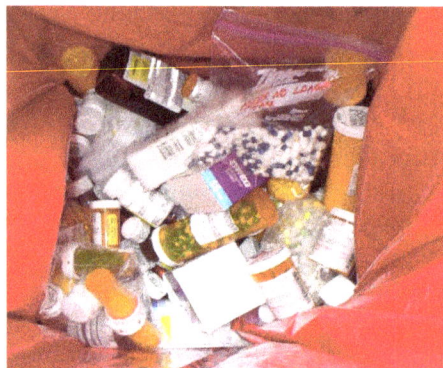

Unused pharmaceuticals collected as part of a university research project into pharmaceuticals waste.

Antibiotics have no effect on viral infections such as the common cold. They are also ineffective against sore throats, which are usually viral and self-resolving. Most cases of bronchitis (90–95%) are viral as well, passing after a few weeks—the use of antibiotics against bronchitis is superfluous and can put the patient at risk of suffering adverse reactions.

Official guidelines by the American Heart Association for dental antibiotic prophylaxis call for the administration of antibiotics to prevent infective endocarditis. Though the current (2007) guidelines dictate more restricted antibiotic use, many dentists and dental patients follow the 1997 guidelines instead, leading to overuse of antibiotics.

Antibiotics in Livestock

There has been massive use of antibiotics in animal husbandry. The most abundant use of antimicrobials worldwide are in livestock; they are typically distributed in animal feed or water for purposes such as disease prevention and growth. Debates have arisen surrounding the extent of the impact of these antibiotics, particularly antimicrobial growth promoters, on human antibiotic resistance. Although some sources believe that there remains a lack of knowledge on which antibiotic use generates the most risk to humans, policies and regulations have been placed to limit any harmful effects.

Antibiotic use in Livestock

Antibiotic use in livestock is the use of antibiotics for any purpose in the husbandry of livestock, which includes treatment when ill (therapeutic), treatment of a batch of animals when at least one is diagnosed as ill (metaphylaxis - similar to the way bacterial meningitis is treated in children), preventative treatment against disease or prophylaxis of infection, but also the use of subtherapeutic doses in animal feed and/or water to promote growth and improve feed efficiency in intensive animal farming outside of Europe, where the last practice has been banned since 2006. This article primarily looks at antibiotic use for growth promotion and the situation in the United States.

Antimicrobials (including antibiotics and antifungals) and other drugs can be used by veterinarians and livestock owners to increase the growth rates of livestock, poultry, and other farmed animals, although these pharmaceuticals do not always have to be administered by a veterinarian. It is important to note the use of antibiotics for growth promotion purposes has been banned in Europe since 2006.

There are also global concerns over the use of antibiotics for growth promotion or therapy purposes because of the potential for some drugs to enter the human food chain, despite rigorous withdrawal measures and testing exists to prevent this, and the problem of increasing antibiotic resistance in animals and a potential although largely unproven link to antibiotic-resistant infections in humans, and what some consider antibiotic misuse. Other drugs may be used only under strict limits, and some organizations and authorities seek to further restrict the use of some or all drugs in animals. Other authorities, such as the World Organization for Animal Health, say that "Without antibiotics there would supply problems of animal protein for the human population".

However, in 2013 the CDC finalized and released a report detailing antibiotic resistance and classified the top 18 resistant bacterium as either being urgent, serious or concerning threats (CDC). Of those organisms, three (CDIFF, CRE and Neisseria gonorrhoeae) have been classified as urgent threats and require more monitoring and prevention (CDC). In the US alone, more than 2 million people are diagnosed with antibiotic resistant infections and over 23,000 die per year due to resistant infections (CDC).

History of the Practice

In 1910 in the United States, a meat shortage resulted in protests and boycotts. After this and other shortages, the public demanded government research into stabilization of food supplies. Since the

1900s, livestock production on United States farms has had to rear larger quantities of animals over a short period of time to meet new consumer demands. Factory farming or the use of high intensity feedlots originated in the late 19th century when advances in technology and science allowed for mass production of livestock. Global agriculture production doubled four times within 1820 and 1975, feeding one billion in 1800 and up to 6.5 billion in 2002. Along with the new large animal densities came the threat of disease, therefore requiring a greater disease control of these animals. In 1950, a group of United States scientists found that adding antibiotics to animal feed increases the growth rate of livestock. American Cyanamid published research establishing the practice.

By 2001 this practice had grown so much that a report by the Union of Concerned Scientists found that nearly 90% of the total use of antimicrobials in the United States was for non-therapeutic purposes in agricultural production.

Antibiotics have an appropriate place in the humane care of illness in livestock, when they reduce the suffering of a sick animal or control the spread of the illness to nearby animals. Thus, ideas that they should *never* be used in livestock husbandry are misguided. Instead, the goal is to prevent the allowance of preventive use from being distorted into routine use, constituting overuse.

Drugs and Growth Stimulation

Certain antibiotics, when given in low, sub-therapeutic doses, are known to improve feed conversion efficiency (more output, such as muscle or milk, for a given amount of feed) and/or may promote greater growth, most likely by affecting gut flora.

Antibiotic Growth Promoters used in Livestock Production		
drug	class	effect
Bambermycin		increase feed conversion ratio and weight gain in chickens, beef cattle, swine, and turkeys.
Lasalocid	Ionophore	increase feed conversion ratio and weight gain in beef cattle.
Monensin	Ionophore	increase feed conversion ratio and weight gain in beef cattle and sheep; promotes proficient milk production in dairy cows.
Salinomycin	Ionophore	increase feed conversion ratio and weight gain.
Virginiamycin	peptide	increase feed conversion ratio and weight gain in chickens, swine, turkeys, and beef cattle.
Bacitracin	peptide	increase weight gain and feed conversion ratio in chickens, turkeys, beef cattle, and swine; promotes egg production in chickens.
Carbadox		increase feed conversion ratio and weight gain in swine.
Laidlomycin		increase feed conversion ratio and weight gain in beef cattle.
Lincomycin		increase feed conversion ratio and weight gain in chickens and swine.
Neomycin/oxytetracyclinee		increase weight gain and feed conversion ratio in chickens, turkeys, swine, and beef cattle.
Penicillin		increase feed conversion ratio and weight gain in chickens, turkeys, and swine.
Roxarsone		increase feed conversion ratio and weight gain in chickens and turkeys.
Tylosin		increase feed conversion ratio and weight gain in chickens and swine.

Use in Different Livestock

In Swine Production

The use of antibiotics to increase the growth of pigs is most studied of all livestock. This use for growth rather than disease prevention is referred to as subtherapeutic antibiotic use. Studies have shown that administering low doses of antibiotics in livestock feed improves growth rate, reduces mortality and morbidity, and improves reproductive performance. It is estimated that over one-half of the antibiotics produced and sold in the United States is used as a feed additive. Although it is still not completely understood why and how antibiotics increase the growth rate of pigs, possibilities include metabolic effects, disease control effects, and nutritional effects. While subtherapeutic use has many benefits for raising swine, there is growing concern that this practice leads to increased antibiotic resistance in bacteria. Antibiotic resistance occurs when bacteria are resistant to one or more microbial agents that are usually used to treat infection. There are three stages in the possible emergence and continuation of antibiotic resistance: genetic change, antibiotic selection, and spread of antibiotic resistance.

In Production of Other Livestock

Organic beef comes from cattle who have not been fed antibiotics.

Regulatory Context

The use of drugs in food animals is regulated in nearly all countries. Historically, this has been to prevent alteration or contamination of meat, milk, eggs and other products with toxins that are harmful to humans. Treating a sick animal with drugs may lead to some of those drugs remaining in the animal when it is slaughtered or milked. Scientific experiments provide data that shows how long a drug is present in the body of an animal and what the animal's body does to the drug. Of particular concern are drugs that may be passed into milk or eggs. By the use of 'drug withdrawal periods' before slaughter or the use of milk or eggs from treated animals, veterinarians and animal owners ensure that the meat, milk and eggs is free of contamination.

These restrictions include not only poisons or drugs (such as penicillin) which may result in allergic reactions but also contaminants which may cause cancer. It is illegal in the USA to administer drugs or feed substances to animals if they have been shown to cause cancer.

One of the main restrictions is the amount that is administered to animals in the industry. These drugs should be administered to healthy livestock at a low concentration of 200 g per ton of feed.

The amount distributed is also altered throughout the lifespan of livestock in order to meet specific growth needs.

Legality of the use of specific drugs in animal medicine varies according to location.

Just as in human medicine, some drugs are available over the counter and others are restricted to use only on the prescription of a veterinary physician. In the USA, the Food and Drug Administration (FDA) requires specific labels on all drugs, giving directions on the use of the drug. For animals, this includes the species, dose, reason for giving the drug (indication) and the required withdrawal period, if any. Federal law requires laypersons to use drugs only in the manner listed. Veterinarians who have examined an animal or a herd of animals may issue a replacement label, giving new directions, based on their medical knowledge. It is illegal in the USA for any layperson to administer any drug to a food animal in a way not specific to the drug label. Over-the-counter drugs which may be used by laypersons include anti-parasite drugs (including fly sprays) and antimicrobials. These drugs can be applied as sprays, creams, injections, oral pills or fluids, or as a feed additive, depending on the drug and the label.

In December 2013, the FDA updated its regulations to try to begin reducing use of antibiotics for growth enhancement. Significant lobbying comes from all directions, from those against tighter regulation to those who complain it doesn't go far enough.

Currently few policies, regulations and laws exist that promote limitation of antibiotic use on factory farms. In addition, few policies are being created that call for this decrease in antibiotic use. However, numerous state senators and members of congress showed support for the Preventing Antibiotic Resistance Act of 2015 (PARA) and the Preservation of Antibiotics for Medical Treatment Act of 2013 (PAMTA). These acts proposed amendments be made to the Federal Food, Drug and Cosmetic Act which would limit and preserve the use of antibiotics for medically necessary situations. Both of these bills died in Congress in 2015.

Administering Drugs

Drugs can be administered to animals in a variety of means, just as with humans. Among these are topical (on the skin), by injection (including intravenous, subcutaneous, subcutaneous implants, intramuscular and intraperitoneal), and orally. Oral drugs can be in pill or liquid form, or can be given by mixing with feed or drinking water. The appropriate route for treatment depends on the specific case and can vary by: illness, severity of illness, selected drug, age or condition of the animal, species of the animal, type of housing and other factors. For animals that are not regularly fed a concentrated feed or which can be handled repeatedly, a slow-release injection might be the most appropriate. Some drugs are not available or appropriate in this form and should be delivered orally. For animals that are fed regularly (rather than grazing freely) or that can not be easily handled, the most appropriate means of administering the drug may be to include the drug in feed or water. This eliminates the stress of daily (or more frequent) handling of animals, which can make the animals more ill. Poultry are most commonly medicated in this fashion, as they are easily stressed to the point of dying. Administering the drug by feed also prevents injection wounds in animals.

The timely administration of drugs is key to preventing animal suffering and economic loss to the farmer. Animals which are ill can infect other animals, and may become so ill that they can not be sold. A variety of techniques are used to monitor animals for illness so that they can be treated appropriately. Stress reduction, adequate nutrition, shelter, and quarantine of incoming stock are all important factors to promote growth and reduce illness and the need for active treatment. The age and status of an animal is also important in determining correct treatment – a young animal or pregnant animal is at greater risk and are treated more aggressively than an older animal. Specifically in calves, the period in which they begin to separate from their mothers generates stress and makes them more susceptible to catching an infection like pneumonia. Antibiotics are commonly administered in the calves' feed during this time to fight the possibility of stress-induced infections. Feed antibiotics are also used to prevent illnesses in calves caused by liver abscesses that develop during their last stages of growth.

Use by Country

European Union

The European Union (EU) in 1999 implemented an antibiotic resistance monitoring program and phase out plan for all antibiotic use by 2006. Although the European Union banned the use of antibiotics as growth agents from 2006, its use has not changed much until recently. In Germany, 1,734 tons of antimicrobial agents were used for animals in 2011 compared with 800 tons for humans. On the other hand, Sweden banned their use in 1986 and Denmark started cutting down drastically in 1994, so that its use is now 60% less. In the Netherlands, the use of antibiotics to treat diseases increased after the ban on its use for growth purposes in 2006. In 2011, the EU voted to ban the prophylactic use of antibiotics, alarmed at signs that the overuse of antibiotics is blunting their use for humans.

United States

In 2011, a total of 13.6 million kilograms of antimicrobials were sold for use in food-producing animals in the United States, which represents 80% of all antibiotics sold or distributed in the United States. Of the antibiotics given to animals fom 2009 through 2013, just above 60% distributed for food animal use are "medically-important" drugs, that are also used in humans. The rest are drug classes like ionophores which are not used in human medicine. Due to concerns about the overuse of antibiotics in food-producing animals, the U.S. Food & Drug Administration has implemented new industry guidelines that will restrict the use of medically-important drugs to uses "that are considered necessary for assuring animal health" and will require veterinary oversight. The food animal and veterinary pharmaceutical industries will need to phase out medically important antimicrobial use by January 1, 2017.

China

China produces and consumes the most antibiotics of all countries.

Antibiotic use has been measured by checking the water near factory farms in China. Measurements have also been taken from animal dung.

Half of the antibiotics manufactured in China are used in the production of livestock.

It was calculated that 38.5 million kg (or 84.9 million lbs) of antibiotics were used in China's swine and poultry production in 2012.

India

In 2012 India manufactured about a third of the total amount of antibiotics in the world.

Brazil

Brazil is the world's largest exporter of beef and the government regulates antibiotic use in the cattle production industry.

Concerns about Antibiotic Resistance

More recently, there has been increased concern about the use of anti-microbials in animals (including pets, livestock, and companion animals) contributing to the rise in antibiotic resistant infections in humans. The use of antimicrobials has been linked to the rise of resistance in every drug and species where it has been studied, including humans and livestock. However, the role of antibiotic use in food animals – in contrast to the use of antibiotics in humans – in the rise of resistant infections in humans is in dispute. The use of antimicrobials in various forms is widespread throughout animal industry, and is presented as key to preventing animal suffering and economic loss. It is linked by some activist groups to animal welfare concern, large scale commercial agriculture, international food trade, agricultural protectionist laws, environmental protection (including climate change) and other topics, which make the aims of some groups on both sides of the debate difficult to untangle.

Around 70% of all antibiotics administered are used for livestock. Most of the drugs that are given to livestock are misused and incorporated into their diets daily for the purpose of weight gain or to treat illnesses. The overuse of the antibiotic in livestock is harmful to humans because it creates an antibiotic resistant bacteria that can be transferred through several different ways such as: raw meats, consumption of meats, or it can also be airborne. Waste from food-producing animals can also contain antibiotic-resistant bacteria and is sometimes stored in lagoons. This waste is often sprayed as fertilizer and can thus contaminate crops and water with the antibiotic-resistant bacteria. Antibiotic resistance is harmful to humans because it makes them resistant to certain types of drugs for different diseases, and makes it harder for them to fight off infections.

The World Health Organization has published a list of Critically Important Antimicrobials for Human Medicine with the intent that it be used "as a reference to help formulate and prioritize risk assessment and risk management strategies for containing antimicrobial resistance due to human and non-human antimicrobial use."

Positions of Advocates for Restricting Antibiotic Use

The practice of using antibiotics for growth stimulation is problematic for these reasons:

- it is the largest use of antimicrobials worldwide

- subtherapeutic use of antibiotics results in bacterial resistance

- every important class of antibiotics are being used in this way, making every class less effective

- the bacteria being changed harm humans

Donald Kennedy, former director of the United States Food and Drug Administration, has said "There's no question that routinely administering non-therapeutic doses of antibiotics to food animals contributes to antibiotic resistance." David Aaron Kessler, another former director of the FDA, said that "We have more than enough scientific evidence to justify curbing the rampant use of antibiotics for livestock, yet the food and drug industries are not only fighting proposed legislation to reduce these practices, they also oppose collecting the data."

In 2013 the United States Centers for Disease Control and Prevention (CDC) published a white paper discussing antibiotic resistance threats in the US and calling for "improved use of antibiotics" among other measures to contain the threat to human health. The CDC asked leaders in agriculture, healthcare, and other disciplines to work together to combat the issue of increasing antibiotic resistance.

Some scientists have said that "all therapeutic antimicrobial agents should be available only by prescription for human and veterinary use."

The Pew Charitable Trusts have stated that "hundreds of scientific studies conducted over four decades demonstrate that feeding low doses of antibiotics to livestock breeds antibiotic-resistant superbugs that can infect people. The FDA, the U.S. Department of Agriculture and the Centers for Disease Control and Prevention all testified before Congress that there is a definitive link between the routine, non-therapeutic use of antibiotics in food animal production and the challenge of antibiotic resistance in humans."

Moderate Positions

The World Organisation for Animal Health has acknowledged the need to protect antibiotics but argued against a total ban on antibiotic use in animal production.

Positions of Advocates for Status Quo

In 2011 the National Pork Producers Council, an American trade association, has said "Not only is there no scientific study linking antibiotic use in food animals to antibiotic resistance in humans, as the U.S. pork industry has continually pointed out, but there isn't even adequate data to conduct a study." The statement contradicts scientific consensus, and was issued in response to a United States Government Accountability Office report that asserts "antibiotic use in food animals contributes to the emergence of resistant bacteria that may affect humans".

The National Pork Board, a Government-owned corporation of the United States, has said that "the vast majority of producers use (antibiotics) appropriately."

Effects of Restricting Antibiotic Use

When government regulation restricts use of antibiotics the negative economic impact is not often considered.

Regulation of antibiotics in livestock production would affect the business models of corporations including Tyson Foods, Cargill, and Hormel.

Difficulties with Determining Relevant Facts

It is difficult to set up a comprehensive surveillance system for measuring rates of change in antibiotic resistance. The US Government Accountability Office published a report in 2011 stating that government and commercial agencies had not been collecting sufficient data to make a decision about best practices.

Currently there is no regulatory agency in the United States that systematically collects detailed data on antibiotic use in humans and animals. It is not clear which antibiotics are prescribed for which purpose and at what time. Furthermore, the world has no surveillance infrastructure to monitor emerging antibiotic resistance threats. Because of these issues, it is difficult to quantify antibiotic resistance, to regulate antibiotic prescribing practices, and to detect and respond to rising threats.

Specific Resistance that has been Identified and Human Impact

There have been many studies that document antibiotic resistant bacteria in livestock, though the impact of the different bacteria in humans is still undergoing research. At this time, the most well-documented impact on humans is foodborne gastrointestinal illness. In most cases, these illnesses are mild and do not require antibiotics; though if the infectious bacteria is drug-resistant, research has shown that these bacteria have increased virulence (ability to cause disease), leading to prolonged illness. Furthermore, in approximately 10% of cases, the disease becomes severe, requiring more advanced treatments. These treatments can take the form of intravenous antibiotics, supportive care for blood infections, and hospital stays, leading to higher costs and greater morbidity with a trend toward higher mortality. Severe disease with this outcome is more common with drug-resistant bacteria. Though all people are susceptible, populations shown to be at higher risk for severe disease include children, the elderly, and those with chronic disease.

Over the past 20 years, the most common drug-resistant foodborne bacteria in industrialized countries have been non-typhoidal salmonella and campylobacter. Research has consistently shown the main contributing factors are bacteria sourced in livestock. One example of this was a 1998 outbreak of multidrug-resistant salmonella in Denmark linked back to two Danish swine herds. Coupled with the discovery of this link, there have been improved monitoring systems that have helped to quantify the impact. In the United States, it is estimated that there are approximately 400,000 cases and over 35,000 hospitalizations per year attributable to increasing resistant strains of salmonella and campylobacter. In terms of financial impact in the US, the treatment of non-typhoidal salmonella infections alone is now estimated to cost $365 million per year. In light of this, in its inaugural 2013 report on antibiotic resistance threats in the United States, the CDC identified resistant non-typhoidal salmonella and campylobacter as "serious threats" and called for improved surveillance and intervention in food production moving forward.

There are other bacteria as well, where research is evolving and revealing that bacterial resistance acquired through use in livestock may be contributing to disease in humans. Examples of these include Enterococcus, E. coli 0157 and Staphylococcus Aureus. In the case of foodborne illness from E.coli, though it is still not typically treated with antibiotics because of associated risk of renal failure, increasing rates of antibiotic resistant infections have been correlated with increasing virulence of the bacteria. In the case of enterococcus and staphylococcus aureus, resistant forms of both of these bacteria have resulted in greatly increasing morbidity and mortality in the US. At this point, there have been studies, though a limited number, that definitively link antibiotic use in food production to these resistance patterns in humans and further research will help to further characterize this relationship.

Mechanisms for Transfer to Humans

Humans can be exposed to antibiotic-resistant bacteria by ingesting them through the food supply. Dairy products, ground beef and poultry are the most common foods harboring these pathogens. There is evidence that a large proportion of resistant *E. coli* isolates causing blood stream infections in people are from livestock produced as food.

When manure from antibiotic-fed swine is used as fertilizer elsewhere, the manure may be contaminated with bacteria which can infect humans.

Studies have also shown that direct contact with livestock can lead to the spread of antibiotic-resistant bacteria from animals to humans.,

Action and Advocacy by Country

Legislation and activism worldwide have aimed at restricting antibiotic use in livestock.

European Union

On 1 January 2006 the European Union banned the non-medicinal use of antibiotics in livestock production.

United States

Some grocery stores have policies about voluntarily not selling meat produced by using antibiotics to stimulate growth. In 2012 in the United States advocacy organization Consumers Union organized a petition asking the store Trader Joe's to discontinue the sale of meat produced with antibiotics.

The U.S. Animal Drug User Fee Act was passed by Congress in 2008 and requires that drug manufacturers report all sales of antibiotics into the food animal production industry.

Some proposed legislation in the US has failed to be adopted. The Animal Drug and Animal Generic Drug User Fee Reauthorization Act of 2013 proposes other regulation.

In the United States the danger of emergence of antibiotic-resistant bacterial strains due to wide use of antibiotics to promote weight gain in livestock was determined by the United States Food and Drug Administration in 1977, but nothing effective was done to prevent the practice. In March, 2012 the United States District Court for the Southern District of New York, ruling in an action brought by the Natural Resources Defense Council and others, ordered the FDA to revoke approvals for the use of antibiotics in livestock which violated FDA regulations. On 11 April 2012 the FDA announced a program to phase out unsupervised use of drugs as feed additives and, on a voluntary basis, convert approved uses for antibiotics to therapeutic use only, requiring veterinarian supervision of their use and a prescription.

In response to consumer concerns about the use of antibiotics in poultry, in 2007, Perdue removed all human antibiotics from its feed and launched the Harvestland brand, under which it sold products that met the requirements for an "antibiotic-free" label. By 2014, Perdue had also phased out ionophores (antibiotics used in animals to lower production costs by promoting growth, and preventing disease) from its hatchery and began using the "antibiotic free" labels on its Harvestland, Simply Smart and Perfect Portions products. By 2015, 52% of the company's chickens were raised without the use of any type of antibiotics.

In 1970 the FDA started recommending that antibiotic use in livestock be limited but set no actual regulations governing this recommendation. Further, in 2004 the Government Accountability Office (GAO) heavily critiqued the FDA for not collecting enough information and data on antibiotic use in factory farms. From this the GAO concluded that the FDA does not have enough information to create effective policy changes regarding antibiotic use. In response to this the FDA insisted that more research was being conducted and voluntary efforts within the industry would solve the problem of antibiotic resistance.

China

In 2012 an American newspaper described the Chinese government's regulation of antibiotics in livestock production as "weak".

India

In 2011 the Indian government proposed a "National policy for containment of antimicrobial resistance". Other policies set schedules for requiring that food producing animals not be given antibiotics for a certain amount of time before their food goes to market. A study released by Centre for Science and Environment (CSE) on 30 July 2014 found antibiotic residues in chicken. This study claims that Indians are developing resistance to antibiotics — and hence falling prey to a host of otherwise curable ailments. Some of this resistance might be due to large-scale unregulated use of antibiotics in the poultry industry. CSE finds that India has not set any limits for antibiotic residues in chicken and says that India will have to implement a comprehensive set of regulations including banning of antibiotic use as growth promoters in the poultry industry. Not doing this will put lives of people at risk.

Brazil

Antibiotic resistant bacteria have been found in Brazilian cattle.

South Korea

In 1998 some researchers reported use in livestock production was a factor in the high prevalence of antibiotic resistant bacteria in Korea. In 2007 *The Korea Times* noted that Korea has relatively high usage of antibiotics in livestock production. In 2011 the Korean government banned the use of antibiotics as growth promoters in livestock.

New Zealand

In 1999 the New Zealand government issued a statement that they would not then ban the use of antibiotics in livestock production. In 2007 ABC Online reported on antibiotic use in chicken production in New Zealand.

Research into Alternatives

Increasing concern due to the emergence of antibiotic resistant bacteria has led researchers to look for alternatives to using antibiotics in livestock.

Probiotics, cultures of a single bacteria strain or mixture of different strains, are being studied in livestock as a production enhancer.

Prebiotics are non-digestible carbohydrates. The carbohydrates are mainly made up of oligosaccharides which are short chains of monosaccharides. The two most commonly studied prebiotics are fructooligosaccharides (FOS) and mannanoligosaccharides (MOS). FOS has been studied for use in chicken feed. MOS works as a competitive binding site, as bacteria bind to it rather than the intestine and are carried out.

Bacteriophages are able to infect most bacteria and are easily found in most environments colonized by bacteria, and have been studied as well.

In another study it was found that using probiotics, competitive exclusion, enzymes, immunomodulators and organic acids prevents the spread of bacteria and can all be used in place of antibiotics. Another research team was able to use bacteriocins, antimicrobial peptides and bacteriophages in the control of bacterial infections. While further research is needed in this field, alternative methods have been identified in effectively controlling bacterial infections in animals. All of the alternative methods listed pose no known threat to human health and all can lead the elimination of antibiotics in factory farms. With further research it is highly likely that a cost effective and health effective alternative could and will be found.

References

- Chemical Analysis of Antibiotic Residues in Food. (PDF). John Wiley & Sons, Inc. 2012. pp. 1–60. ISBN 9781449614591.

- "Antimicrobial resistance: global report on surveillance" (PDF). The World Health Organization. April 2014. ISBN 978 92 4 156474 8. Retrieved 13 June 2016.

- Gualerzi, Claudio O.; Brandi, Letizia; Fabbretti, Attilio; Pon, Cynthia L. (2013-12-04). Antibiotics: Targets, Mechanisms and Resistance. John Wiley & Sons. p. 1. ISBN 9783527333059.

- Dyer, Betsey Dexter (2003). "Chapter 9, Pathogens". A Field Guide To Bacteria. Cornell University Press. ISBN 978-0-8014-8854-2.

- Marino PL (2007). "Antimicrobial therapy". The ICU book. Hagerstown, MD: Lippincott Williams & Wilkins. p. 817. ISBN 978-0-7817-4802-5.

- Miller, AA (2011). Miller, PF, ed. Emerging Trends in Antibacterial Discovery: Answering the Call to Arms. Caister Academic Press. ISBN 978-1-904455-89-9.

- Neonatal Formulary: Drug Use in Pregnancy and the First Year of Life (7 ed.). John Wiley & Sons. 2014. p. 162. ISBN 9781118819517.

- Hamilton, Richart (2015). Tarascon Pocket Pharmacopoeia 2015 Deluxe Lab-Coat Edition. Jones & Bartlett Learning. p. 108. ISBN 9781284057560.

- (February 8, 2005) "Osteomyelitis", in Kahn, Cynthia M., Line, Scott, Aiello, Susan E. (ed.): The Merck Veterinary Manual, 9th ed., John Wiley & Sons. ISBN 0-911910-50-6. Retrieved 14 December 2007.

- (February 8, 2005) "Toxoplasmosis: Introduction", in Kahn, Cynthia M., Line, Scott, Aiello, Susan E. (ed.):

The Merck Veterinary Manual, 9th ed., John Wiley & Sons. ISBN 0-911910-50-6. Retrieved 14 December 2007.

- Koplow, David A. (2003). Smallpox: the fight to eradicate a global scourge. Berkeley: University of California Press. ISBN 0-520-24220-3.

- Offit PA (2007). Vaccinated: One Man's Quest to Defeat the World's Deadliest Diseases. Washington, DC: Smithsonian. ISBN 0-06-122796-X.

- Plotkin, Stanley A. (2006). Mass Vaccination: Global Aspects - Progress and Obstacles (Current Topics in Microbiology & Immunology). Springer-Verlag Berlin and Heidelberg GmbH & Co. K. ISBN 978-3-540-29382-8.

- McAuliffe et al. "The New Phage Biology: From Genomics to Applications" (introduction) in Mc Grath, S. and van Sinderen, D. (eds.) Bacteriophage: Genetics and Molecular Biology Caister Academic Press ISBN 978-1-904455-14-1

- Mc Grath S and van Sinderen D (editors). (2007). Bacteriophage: Genetics and Molecular Biology (1st ed.). Caister Academic Press. ISBN 978-1-904455-14-1.

- Abedon ST (2012). Salutary contributions of viruses to medicine and public health. In: Witzany G (ed). Viruses: Essential Agents of Life. Springer. 389-405. ISBN 978-94-007-4898-9.

- Brüssow, H 2007. Phage Therapy: The Western Perspective. in S. McGrath and D. van Sinderen (eds.) Bacteriophage: Genetics and Molecular Biology, Caister Academic Press, Norfolk, UK. ISBN 978-1-904455-14-1

- Kuchment, Anna (2011), The Forgotten Cure: The Past and Future of Phage Therapy, Springer, ISBN 978-1-4614-0250-3

- Häusler, Thomas (2006), Virus vs. Superbug: A solution to the antibiotic crisis?, Macmillan, p. 48, ISBN 978-0-230-55193-0

4

Microbes and its Types

Microbes are organisms which are either single celled or are multicellular. They are diverse and include bacteria, archaea and protozoa. Some of the types discussed within this chapter are methicillin-resistant Staphylococcus aureus, Salmonella, Escherichia coli, Campylobacter, Streptococcus pyogenes etc. Microbes are best understood in confluence with the major topics listed in the following text.

Virus

A virus is a small infectious agent that replicates only inside the living cells of other organisms. Viruses can infect all types of life forms, from animals and plants to microorganisms, including bacteria and archaea.

Since Dmitri Ivanovsky's 1892 article describing a non-bacterial pathogen infecting tobacco plants, and the discovery of the tobacco mosaic virus by Martinus Beijerinck in 1898, about 5,000 virus species have been described in detail, although there are millions of types. Viruses are found in almost every ecosystem on Earth and are the most abundant type of biological entity. The study of viruses is known as virology, a sub-speciality of microbiology.

While not inside an infected cell or in the process of infecting a cell, viruses exist in the form of independent particles. These viral particles, also known as virions, consist of two or three parts: (i) the genetic material made from either DNA or RNA, long molecules that carry genetic information; (ii) a protein coat, called the capsid, which surrounds and protects the genetic material; and in some cases (iii) an envelope of lipids that surrounds the protein coat when they are outside a cell. The shapes of these virus particles range from simple helical and icosahedral forms for some virus species to more complex structures for others. Most virus species have virions that are too small to be seen with an optical microscope. The average virion is about one one-hundredth the size of the average bacterium.

The origins of viruses in the evolutionary history of life are unclear: some may have evolved from plasmids—pieces of DNA that can move between cells—while others may have evolved from bacteria. In evolution, viruses are an important means of horizontal gene transfer, which increases genetic diversity. Viruses are considered by some to be a life form, because they carry genetic material, reproduce, and evolve through natural selection. However they lack key characteristics (such as cell structure) that are generally considered necessary to count as life. Because they possess some but not all such qualities, viruses have been described as "organisms at the edge of life", and as replicators.

Viruses spread in many ways; viruses in plants are often transmitted from plant to plant by insects that feed on plant sap, such as aphids; viruses in animals can be carried by blood-sucking insects.

These disease-bearing organisms are known as vectors. Influenza viruses are spread by coughing and sneezing. Norovirus and rotavirus, common causes of viral gastroenteritis, are transmitted by the faecal–oral route and are passed from person to person by contact, entering the body in food or water. HIV is one of several viruses transmitted through sexual contact and by exposure to infected blood. The range of host cells that a virus can infect is called its "host range". This can be narrow, meaning a virus is capable of infecting few species, or broad, meaning it is capable of infecting many.

Viral infections in animals provoke an immune response that usually eliminates the infecting virus. Immune responses can also be produced by vaccines, which confer an artificially acquired immunity to the specific viral infection. However, some viruses including those that cause AIDS and viral hepatitis evade these immune responses and result in chronic infections. Antibiotics have no effect on viruses, but several antiviral drugs have been developed.

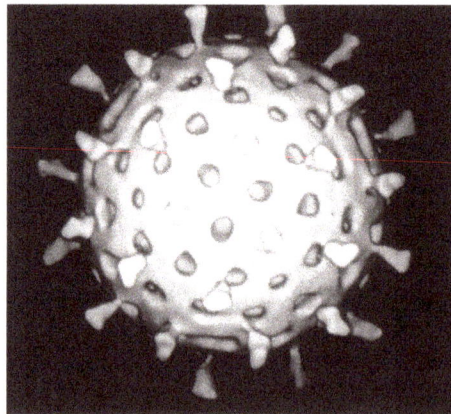

Rotavirus

History

Martinus Beijerinck in his laboratory in 1921

Pasteur was unable to find a causative agent for rabies and speculated about a pathogen too small to be detected using a microscope. In 1884, the French microbiologist Charles Chamberland invented a filter (known today as the Chamberland filter or the Pasteur-Chamberland filter) with pores smaller than bacteria. Thus, he could pass a solution containing bacteria through the filter and completely remove them from the solution. In 1892, the Russian biologist Dmitri Ivanovsky used this filter to study what is now known as the tobacco mosaic virus. His experiments showed that crushed leaf extracts from infected tobacco plants remain infectious after filtration. Ivanovsky suggested the infection might be caused by a toxin produced by bacteria, but did not pursue the idea. At the time it was thought that all infectious agents could be retained by filters and grown on a nutrient medium – this was part of the germ theory of disease. In 1898, the Dutch microbiologist Martinus Beijerinck repeated the experiments and became convinced that the filtered solution contained a new form of infectious agent. He observed that the agent multiplied only in cells that were dividing, but as his experiments did not show that it was made of particles, he called it a *contagium vivum fluidum* (soluble living germ) and re-introduced the word *virus*. Beijerinck maintained that viruses were liquid in nature, a theory later discredited by Wendell Stanley, who proved they were particulate. In the same year Friedrich Loeffler and Paul Frosch passed the first animal virus – agent of foot-and-mouth disease (aphthovirus) – through a similar filter.

In the early 20th century, the English bacteriologist Frederick Twort discovered a group of viruses that infect bacteria, now called bacteriophages (or commonly *phages*), and the French-Canadian microbiologist Félix d'Herelle described viruses that, when added to bacteria on an agar plate, would produce areas of dead bacteria. He accurately diluted a suspension of these viruses and discovered that the highest dilutions (lowest virus concentrations), rather than killing all the bacteria, formed discrete areas of dead organisms. Counting these areas and multiplying by the dilution factor allowed him to calculate the number of viruses in the original suspension. Phages were heralded as a potential treatment for diseases such as typhoid and cholera, but their promise was forgotten with the development of penicillin. The study of phages provided insights into the switching on and off of genes, and a useful mechanism for introducing foreign genes into bacteria.

By the end of the 19th century, viruses were defined in terms of their infectivity, their ability to be filtered, and their requirement for living hosts. Viruses had been grown only in plants and animals. In 1906, Ross Granville Harrison invented a method for growing tissue in lymph, and, in 1913, E. Steinhardt, C. Israeli, and R. A. Lambert used this method to grow vaccinia virus in fragments of guinea pig corneal tissue. In 1928, H. B. Maitland and M. C. Maitland grew vaccinia virus in suspensions of minced hens' kidneys. Their method was not widely adopted until the 1950s, when poliovirus was grown on a large scale for vaccine production.

Another breakthrough came in 1931, when the American pathologist Ernest William Goodpasture and Alice Miles Woodruff grew influenza and several other viruses in fertilised chickens' eggs. In 1949, John Franklin Enders, Thomas Weller, and Frederick Robbins grew polio virus in cultured human embryo cells, the first virus to be grown without using solid animal tissue or eggs. This work enabled Jonas Salk to make an effective polio vaccine.

The first images of viruses were obtained upon the invention of electron microscopy in 1931 by the German engineers Ernst Ruska and Max Knoll. In 1935, American biochemist and virologist Wendell Meredith Stanley examined the tobacco mosaic virus and found it was mostly made of protein. A short time later, this virus was separated into protein and RNA parts. The tobacco mosaic virus was the first to be crystallised and its structure could therefore be elucidated in detail. The first X-ray diffraction pictures of the crystallised virus were obtained by Bernal and Fankuchen in 1941. On the basis of her pictures, Rosalind Franklin discovered the full structure of the virus in 1955. In the same year, Heinz Fraenkel-Conrat and Robley Williams showed that purified tobacco mosaic virus RNA and its protein coat can assemble by themselves to form functional viruses, suggesting that this simple mechanism was probably the means through which viruses were created within their host cells.

The second half of the 20th century was the golden age of virus discovery and most of the over 2,000 recognised species of animal, plant, and bacterial viruses were discovered during these years. In 1957, equine arterivirus and the cause of Bovine virus diarrhoea (a pestivirus) were discovered. In 1963, the hepatitis B virus was discovered by Baruch Blumberg, and in 1965, Howard Temin described the first retrovirus. Reverse transcriptase, the enzyme that retroviruses use to make DNA copies of their RNA, was first described in 1970, independently by Howard Martin Temin and David Baltimore. In 1983 Luc Montagnier's team at the Pasteur Institute in France, first isolated the retrovirus now called HIV. In 1989 Michael Houghton's team at Chiron Corporation discovered Hepatitis C.

Origins

Viruses are found wherever there is life and have probably existed since living cells first evolved. The origin of viruses is unclear because they do not form fossils, so molecular techniques have been used to compare the DNA or RNA of viruses and are a useful means of investigating how they arose. In addition, viral genetic material may occasionally integrate into the germline of the host organisms, by which they can be passed on vertically to the offspring of the host for many generations. This provides an invaluable source of information for paleovirologists to trace back ancient viruses that have existed up to millions of years ago. Currently, there are three main hypotheses that aim to explain the origins of viruses:

Regressive Hypothesis

> Viruses may have once been small cells that parasitised larger cells. Over time, genes not required by their parasitism were lost. The bacteria rickettsia and chlamydia are living cells that, like viruses, can reproduce only inside host cells. They lend support to this hypothesis, as their dependence on parasitism is likely to have caused the loss of genes that enabled them to survive outside a cell. This is also called the *degeneracy hypothesis*, or *reduction hypothesis*.

Cellular Origin Hypothesis

> Some viruses may have evolved from bits of DNA or RNA that "escaped" from the genes of a larger organism. The escaped DNA could have come from plasmids (pieces of naked DNA that can move *between* cells) or transposons (molecules of DNA that replicate and move around to different positions *within* the genes of the cell). Once called "jumping genes", transposons are examples of mobile genetic elements and could be the origin of some viruses. They were discovered in maize by Barbara McClintock in 1950. This is sometimes called the *vagrancy hypothesis*, or the *escape hypothesis*.

Co-evolution Hypothesis

> This is also called the *virus-first hypothesis* and proposes that viruses may have evolved from complex molecules of protein and nucleic acid at the same time as cells first appeared on Earth and would have been dependent on cellular life for billions of years. Viroids are molecules of RNA that are not classified as viruses because they lack a protein coat. However, they have characteristics that are common to several viruses and are often called subviral agents. Viroids are important pathogens of plants. They do not code for proteins but interact with the host cell and use the host machinery for their replication. The hepatitis delta virus of humans has an RNA genome similar to viroids but has a protein coat derived from hepatitis B virus and cannot produce one of its own. It is, therefore, a defective virus. Although hepatitis delta virus genome may replicate independently once inside a host cell, it requires the help of hepatitis B virus to provide a protein coat so that it can be transmitted to new cells. In similar manner, the sputnik virophage is dependent on mimivirus, which infects the protozoan *Acanthamoeba castellanii*. These viruses, which are dependent on the presence of other virus species in the host cell, are called *satellites* and may represent evolutionary intermediates of viroids and viruses.

In the past, there were problems with all of these hypotheses: the regressive hypothesis did not explain why even the smallest of cellular parasites do not resemble viruses in any way. The escape hypothesis did not explain the complex capsids and other structures on virus particles. The virus-first hypothesis contravened the definition of viruses in that they require host cells. Viruses are now recognised as ancient and as having origins that pre-date the divergence of life into the three domains. This discovery has led modern virologists to reconsider and re-evaluate these three classical hypotheses.

The evidence for an ancestral world of RNA cells and computer analysis of viral and host DNA sequences are giving a better understanding of the evolutionary relationships between different viruses and may help identify the ancestors of modern viruses. To date, such analyses have not proved which of these hypotheses is correct. However, it seems unlikely that all currently known viruses have a common ancestor, and viruses have probably arisen numerous times in the past by one or more mechanisms.

Prions are infectious protein molecules that do not contain DNA or RNA. They can cause infections such as scrapie in sheep, bovine spongiform encephalopathy ("mad cow" disease) in cattle, and chronic wasting disease in deer; in humans, prionic diseases include Kuru, Creutzfeldt–Jakob disease, and Gerstmann–Sträussler–Scheinker syndrome. Although prions are fundamentally different from viruses and viroids, their discovery gives credence to the theory that viruses could have evolved from self-replicating molecules.

Microbiology

Life Properties

Opinions differ on whether viruses are a form of life, or organic structures that interact with living organisms. They have been described as "organisms at the edge of life", since they resemble organisms in that they possess genes, evolve by natural selection, and reproduce by creating multiple copies of themselves through self-assembly. Although they have genes, they do not have a cellular structure, which is often seen as the basic unit of life. Viruses do not have their own metabolism, and require a host cell to make new products. They therefore cannot naturally reproduce outside a host cell – although bacterial species such as rickettsia and chlamydia are considered living organisms despite the same limitation. Accepted forms of life use cell division to reproduce, whereas viruses spontaneously assemble within cells. They differ from autonomous growth of crystals as they inherit genetic mutations while being subject to natural selection. Virus self-assembly within host cells has implications for the study of the origin of life, as it lends further credence to the hypothesis that life could have started as self-assembling organic molecules.

Structure

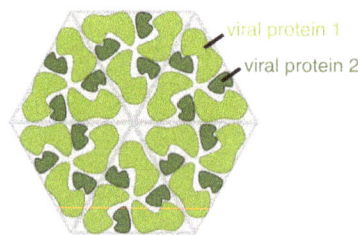

Diagram of how a virus capsid can be constructed using multiple copies of just two protein molecules

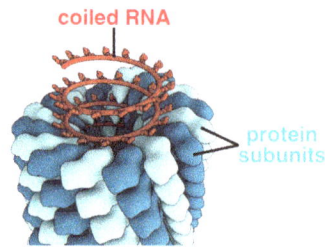

Structure of tobacco mosaic virus: RNA coiled in a helix of repeating protein sub-units

Structure of icosahedral adenovirus. Electron micrograph of with a cartoon to show shape

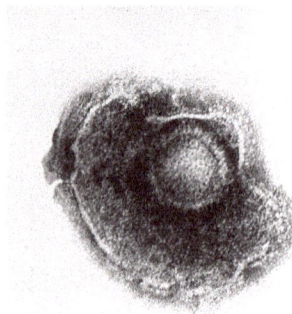

Structure of chickenpox virus. They have a lipid envelope

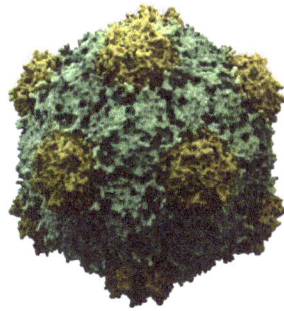

Structure of an icosahedral cowpea mosaic virus

Viruses display a wide diversity of shapes and sizes, called *morphologies*. In general, viruses are much smaller than bacteria. Most viruses that have been studied have a diameter between 20 and 300 nanometres. Some filoviruses have a total length of up to 1400 nm; their diameters are only about 80 nm. Most viruses cannot be seen with an optical microscope so scanning and transmission electron microscopes are used to visualise virions. To increase the contrast between viruses and the background, electron-dense "stains" are used. These are solutions of salts of heavy metals, such as tungsten, that scatter the electrons from regions covered with the stain. When virions are coated with stain (positive staining), fine detail is obscured. Negative staining overcomes this problem by staining the background only.

A complete virus particle, known as a virion, consists of nucleic acid surrounded by a protective coat of protein called a capsid. These are formed from identical protein subunits called capsomeres. Viruses can have a lipid "envelope" derived from the host cell membrane. The capsid is made from proteins encoded by the viral genome and its shape serves as the basis for morphological distinction. Virally coded protein subunits will self-assemble to form a capsid, in general requiring the presence of the virus genome. Complex viruses code for proteins that assist in the construction of their capsid. Proteins associated with nucleic acid are known as nucleoproteins, and the association of viral capsid proteins with viral nucleic acid is called a nucleocapsid. The capsid and entire virus structure can be mechanically (physically) probed through atomic force microscopy. In general, there are four main morphological virus types:

Helical

> These viruses are composed of a single type of capsomere stacked around a central axis to form a helical structure, which may have a central cavity, or tube. This arrangement results in rod-shaped or filamentous virions: These can be short and highly rigid, or long and very flexible. The genetic material, in general, single-stranded RNA, but ssDNA in some cases, is bound into the protein helix by interactions between the negatively charged nucleic acid and positive charges on the protein. Overall, the length of a helical capsid is related to the length of the nucleic acid contained within it and the diameter is dependent on the size and arrangement of capsomeres. The well-studied tobacco mosaic virus is an example of a helical virus.

Icosahedral

> Most animal viruses are icosahedral or near-spherical with chiral icosahedral symmetry. A regular icosahedron is the optimum way of forming a closed shell from identical sub-units.

The minimum number of identical capsomeres required is twelve, each composed of five identical sub-units. Many viruses, such as rotavirus, have more than twelve capsomers and appear spherical but they retain this symmetry. Capsomeres at the apices are surrounded by five other capsomeres and are called pentons. Capsomeres on the triangular faces are surrounded by six others and are called hexons. Hexons are in essence flat and pentons, which form the 12 vertices, are curved. The same protein may act as the subunit of both the pentamers and hexamers or they may be composed of different proteins.

Prolate

This is an icosahedron elongated along the fivefold axis and is a common arrangement of the heads of bacteriophages. This structure is composed of a cylinder with a cap at either end.

Envelope

Some species of virus envelop themselves in a modified form of one of the cell membranes, either the outer membrane surrounding an infected host cell or internal membranes such as nuclear membrane or endoplasmic reticulum, thus gaining an outer lipid bilayer known as a viral envelope. This membrane is studded with proteins coded for by the viral genome and host genome; the lipid membrane itself and any carbohydrates present originate entirely from the host. The influenza virus and HIV use this strategy. Most enveloped viruses are dependent on the envelope for their infectivity.

Complex

These viruses possess a capsid that is neither purely helical nor purely icosahedral, and that may possess extra structures such as protein tails or a complex outer wall. Some bacteriophages, such as Enterobacteria phage T4, have a complex structure consisting of an icosahedral head bound to a helical tail, which may have a hexagonal base plate with protruding protein tail fibres. This tail structure acts like a molecular syringe, attaching to the bacterial host and then injecting the viral genome into the cell.

The poxviruses are large, complex viruses that have an unusual morphology. The viral genome is associated with proteins within a central disc structure known as a nucleoid. The nucleoid is surrounded by a membrane and two lateral bodies of unknown function. The virus has an outer envelope with a thick layer of protein studded over its surface. The whole virion is slightly pleiomorphic, ranging from ovoid to brick shape. Mimivirus is one of the largest characterised viruses, with a capsid diameter of 400 nm. Protein filaments measuring 100 nm project from the surface. The capsid appears hexagonal under an electron microscope, therefore the capsid is probably icosahedral. In 2011, researchers discovered the largest then known virus in samples of water collected from the ocean floor off the coast of Las Cruces, Chile. Provisionally named *Megavirus chilensis*, it can be seen with a basic optical microscope. In 2013, the Pandoravirus genus was discovered in Chile and Australia, and has genomes about twice as large as Megavirus and Mimivirus.

Some viruses that infect Archaea have complex structures that are unrelated to any other form of virus, with a wide variety of unusual shapes, ranging from spindle-shaped structures, to viruses that resemble hooked rods, teardrops or even bottles. Other archaeal viruses resemble the tailed bacteriophages, and can have multiple tail structures.

Genome

Genomic diversity among viruses	
Property	Parameters
Nucleic acid	• DNA • RNA • Both DNA and RNA (at different stages in the life cycle)
Shape	• Linear • Circular • Segmented
Strandedness	• Single-stranded • Double-stranded • Double-stranded with regions of single-strandedness
Sense	• Positive sense (+) • Negative sense (−) • Ambisense (+/−)

An enormous variety of genomic structures can be seen among viral species; as a group, they contain more structural genomic diversity than plants, animals, archaea, or bacteria. There are millions of different types of viruses, although only about 5,000 types have been described in detail. As of September 2015, the NCBI Virus genome database has more than 75,000 complete genome sequences. but there are doubtlessly many more to be discovered.

A virus has either a DNA or an RNA genome and is called a DNA virus or an RNA virus, respectively. The vast majority of viruses have RNA genomes. Plant viruses tend to have single-stranded RNA genomes and bacteriophages tend to have double-stranded DNA genomes.

Viral genomes are *circular*, as in the polyomaviruses, or *linear*, as in the adenoviruses. The type of nucleic acid is irrelevant to the shape of the genome. Among RNA viruses and certain DNA viruses, the genome is often divided up into separate parts, in which case it is called segmented. For RNA viruses, each segment often codes for only one protein and they are usually found together in one capsid. However, all segments are not required to be in the same virion for the virus to be infectious, as demonstrated by brome mosaic virus and several other plant viruses.

A viral genome, irrespective of nucleic acid type, is almost always either *single-stranded* or *double-stranded*. Single-stranded genomes consist of an unpaired nucleic acid, analogous to one-half of a ladder split down the middle. Double-stranded genomes consist of two complementary paired nucleic acids, analogous to a ladder. The virus particles of some virus families, such as those belonging to the *Hepadnaviridae*, contain a genome that is partially double-stranded and partially single-stranded.

For most viruses with RNA genomes and some with single-stranded DNA genomes, the single strands are said to be either positive-sense (called the *plus-strand*) or negative-sense (called the *minus-strand*), depending on if they are complementary to the viral messenger RNA (mRNA). Positive-sense viral RNA is in the same sense as viral mRNA and thus at least a part of it can be immediately translated by the host cell. Negative-sense viral RNA is complementary to mRNA and thus must be converted to positive-sense RNA by an RNA-dependent RNA polymerase before translation. DNA nomenclature for viruses with single-sense genomic ssDNA is similar to RNA nomenclature, in that the *template strand* for the viral mRNA is complementary to it (−), and the *coding strand* is a copy of it (+). However, several types of ssDNA and ssRNA viruses have genomes that are ambisense in that transcription can occur off both strands in a double-stranded replicative intermediate. Examples include geminiviruses, which are ssDNA plant viruses and arenaviruses, which are ssRNA viruses of animals.

Genome size varies greatly between species. The smallest viral genomes – the ssDNA circoviruses, family *Circoviridae* – code for only two proteins and have a genome size of only two kilobases; the largest–the pandoraviruses–have genome sizes of around two megabases which code for about 2500 proteins.

In general, RNA viruses have smaller genome sizes than DNA viruses because of a higher error-rate when replicating, and have a maximum upper size limit. Beyond this limit, errors in the genome when replicating render the virus useless or uncompetitive. To compensate for this, RNA viruses often have segmented genomes – the genome is split into smaller molecules – thus reducing the chance that an error in a single-component genome will incapacitate the entire genome. In contrast, DNA viruses generally have larger genomes because of the high fidelity of their replication enzymes. Single-strand DNA viruses are an exception to this rule, however, as mutation rates for these genomes can approach the extreme of the ssRNA virus case.

Genetic Mutation

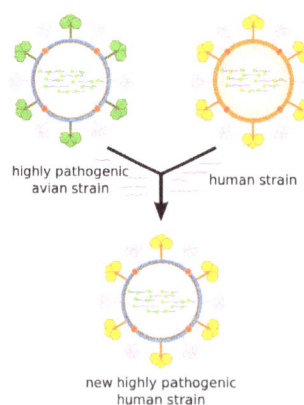

How antigenic shift, or reassortment, can result in novel and highly pathogenic strains of human flu

Viruses undergo genetic change by several mechanisms. These include a process called antigenic drift where individual bases in the DNA or RNA mutate to other bases. Most of these point mutations are "silent" – they do not change the protein that the gene encodes – but others can confer evolutionary advantages such as resistance to antiviral drugs. Antigenic shift occurs when there is a major change in the genome of the virus. This can be a result of recombination or reassortment.

When this happens with influenza viruses, pandemics might result. RNA viruses often exist as quasispecies or swarms of viruses of the same species but with slightly different genome nucleoside sequences. Such quasispecies are a prime target for natural selection.

Segmented genomes confer evolutionary advantages; different strains of a virus with a segmented genome can shuffle and combine genes and produce progeny viruses or (offspring) that have unique characteristics. This is called reassortment or *viral sex.*

Genetic recombination is the process by which a strand of DNA is broken and then joined to the end of a different DNA molecule. This can occur when viruses infect cells simultaneously and studies of viral evolution have shown that recombination has been rampant in the species studied. Recombination is common to both RNA and DNA viruses.

Replication Cycle

Viral populations do not grow through cell division, because they are acellular. Instead, they use the machinery and metabolism of a host cell to produce multiple copies of themselves, and they *assemble* in the cell.

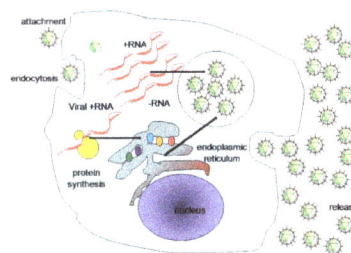

A typical virus replication cycle

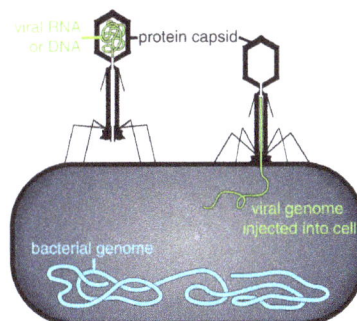

Some bacteriophages inject their genomes into bacterial cells (not to scale)

The life cycle of viruses differs greatly between species but there are six *basic* stages in the life cycle of viruses:

Attachment is a specific binding between viral capsid proteins and specific receptors on the host cellular surface. This specificity determines the host range of a virus. For example, HIV infects a limited range of human leucocytes. This is because its surface protein, gp120, specifically interacts with the CD4 molecule – a chemokine receptor – which is most commonly found on the surface of CD4+ T-Cells. This mechanism has evolved to favour those viruses that infect only cells in which they are capable of replication. Attachment to the receptor can induce the viral envelope protein to undergo changes that results in the fusion of viral and cellular membranes, or changes of non-enveloped virus surface proteins that allow the virus to enter.

Penetration follows attachment: Virions enter the host cell through receptor-mediated endocytosis or membrane fusion. This is often called viral entry. The infection of plant and fungal cells is different from that of animal cells. Plants have a rigid cell wall made of cellulose, and fungi one of chitin, so most viruses can get inside these cells only after trauma to the cell wall. However, nearly all plant viruses (such as tobacco mosaic virus) can also move directly from cell to cell, in the form of single-stranded nucleoprotein complexes, through pores called plasmodesmata. Bacteria, like plants, have strong cell walls that a virus must breach to infect the cell. However, given that bacterial cell walls are much less thick than plant cell walls due to their much smaller size, some viruses have evolved mechanisms that inject their genome into the bacterial cell across the cell wall, while the viral capsid remains outside.

Uncoating is a process in which the viral capsid is removed: This may be by degradation by viral enzymes or host enzymes or by simple dissociation; the end-result is the releasing of the viral genomic nucleic acid.

Replication of viruses involves primarily multiplication of the genome. Replication involves synthesis of viral messenger RNA (mRNA) from "early" genes (with exceptions for positive sense RNA viruses), viral protein synthesis, possible assembly of viral proteins, then viral genome replication mediated by early or regulatory protein expression. This may be followed, for complex viruses with larger genomes, by one or more further rounds of mRNA synthesis: "late" gene expression is, in general, of structural or virion proteins.

Assembly – Following the structure-mediated self-*assembly* of the virus particles, some modification of the proteins often occurs. In viruses such as HIV, this modification (sometimes called maturation) occurs *after* the virus has been released from the host cell.

Release – Viruses can be *released* from the host cell by lysis, a process that kills the cell by bursting its membrane and cell wall if present: This is a feature of many bacterial and some animal viruses. Some viruses undergo a lysogenic cycle where the viral genome is incorporated by genetic recombination into a specific place in the host's chromosome. The viral genome is then known as a "provirus" or, in the case of bacteriophages a "prophage". Whenever the host divides, the viral genome is also replicated. The viral genome is mostly silent within the host. However, at some point, the provirus or prophage may give rise to active virus, which may lyse the host cells. Enveloped viruses (e.g., HIV) typically are released from the host cell by budding. During this process the virus acquires its envelope, which is a modified piece of the host's plasma or other, internal membrane.

The genetic material within virus particles, and the method by which the material is replicated, varies considerably between different types of viruses.

DNA viruses

The genome replication of most DNA viruses takes place in the cell's nucleus. If the cell has the appropriate receptor on its surface, these viruses enter the cell sometimes by direct fusion with the cell membrane (e.g., herpesviruses) or – more usually – by receptor-mediated endocytosis. Most DNA viruses are entirely dependent on the host cell's DNA and RNA synthesising machinery, and RNA processing machinery. However, viruses with larger genomes may encode much of this machinery themselves. In eukaryotes the viral genome must cross the cell's nuclear membrane to access this machinery, while in bacteria it need only enter the cell.

RNA viruses

Replication usually takes place in the cytoplasm. RNA viruses can be placed into four different groups depending on their modes of replication. The polarity (whether or not it can be used directly by ribosomes to make proteins) of single-stranded RNA viruses largely determines the replicative mechanism; the other major criterion is whether the genetic material is single-stranded or double-stranded. All RNA viruses use their own RNA replicase enzymes to create copies of their genomes.

Reverse transcribing viruses

These have ssRNA (*Retroviridae, Metaviridae, Pseudoviridae*) or dsDNA (*Caulimoviridae*, and *Hepadnaviridae*) in their particles. Reverse transcribing viruses with RNA genomes (retroviruses), use a DNA intermediate to replicate, whereas those with DNA genomes (pararetroviruses) use an RNA intermediate during genome replication. Both types use a reverse transcriptase, or RNA-dependent DNA polymerase enzyme, to carry out the nucleic acid conversion. Retroviruses integrate the DNA produced by reverse transcription into the host genome as a provirus as a part of the replication process; pararetroviruses do not, although integrated genome copies of especially plant pararetroviruses can give rise to infectious virus. They are susceptible to antiviral drugs that inhibit the reverse transcriptase enzyme, e.g. zidovudine and lamivudine. An example of the first type is HIV, which is a retrovirus. Examples of the second type are the *Hepadnaviridae*, which includes Hepatitis B virus.

Effects on the Host Cell

The range of structural and biochemical effects that viruses have on the host cell is extensive. These are called *cytopathic effects*. Most virus infections eventually result in the death of the host cell. The causes of death include cell lysis, alterations to the cell's surface membrane and apoptosis. Often cell death is caused by cessation of its normal activities because of suppression by virus-specific proteins, not all of which are components of the virus particle.

Some viruses cause no apparent changes to the infected cell. Cells in which the virus is latent and inactive show few signs of infection and often function normally. This causes persistent infections and the virus is often dormant for many months or years. This is often the case with herpes viruses. Some viruses, such as Epstein–Barr virus, can cause cells to proliferate without causing malignancy, while others, such as papillomaviruses, are established causes of cancer.

Host Range

Viruses are by far the most abundant biological entities on Earth and they outnumber all the others put together. They infect all types of cellular life including animals, plants, bacteria and fungi.

However, different types of viruses can infect only a limited range of hosts and many are species-specific. Some, such as smallpox virus for example, can infect only one species – in this case humans, and are said to have a narrow host range. Other viruses, such as rabies virus, can infect different species of mammals and are said to have a broad range. The viruses that infect plants are harmless to animals, and most viruses that infect other animals are harmless to humans. The host range of some bacteriophages is limited to a single strain of bacteria and they can be used to trace the source of outbreaks of infections by a method called phage typing.

Classification

Classification seeks to describe the diversity of viruses by naming and grouping them on the basis of similarities. In 1962, André Lwoff, Robert Horne, and Paul Tournier were the first to develop a means of virus classification, based on the Linnaean hierarchical system. This system bases classification on phylum, class, order, family, genus, and species. Viruses were grouped according to their shared properties (not those of their hosts) and the type of nucleic acid forming their genomes. Later the International Committee on Taxonomy of Viruses was formed. However, viruses are not classified on the basis of phylum or class, as their small genome size and high rate of mutation makes it difficult to determine their ancestry beyond order. As such, the Baltimore classification is used to supplement the more traditional hierarchy.

ICTV Classification

The International Committee on Taxonomy of Viruses (ICTV) developed the current classification system and wrote guidelines that put a greater weight on certain virus properties to maintain family uniformity. A unified taxonomy (a universal system for classifying viruses) has been established. The 9th lCTV Report defines the concept of the virus species as the lowest taxon (group) in a branching hierarchy of viral taxa. However, at present only a small part of the total diversity of viruses has been studied, with analyses of samples from humans finding that about 20% of the virus sequences recovered have not been seen before, and samples from the environment, such as from seawater and ocean sediments, finding that the large majority of sequences are completely novel.

The general taxonomic structure is as follows:

> Order (-*virales*)
>
> Family (-*viridae*)
>
> Subfamily (-*virinae*)
>
> Genus (-*virus*)
>
> Species (-*virus*)

In the current (2013) ICTV taxonomy, 7 orders have been established, the *Caudovirales*, *Herpesvirales*, *Ligamenvirales*, *Mononegavirales*, *Nidovirales*, *Picornavirales*, and *Tymovirales*. The committee does not formally distinguish between subspecies, strains, and isolates. In total there are 7 orders, 103 families, 22 subfamilies, 455 genera, about 2,827 species and over 4,000 types yet unclassified.

Baltimore Classification

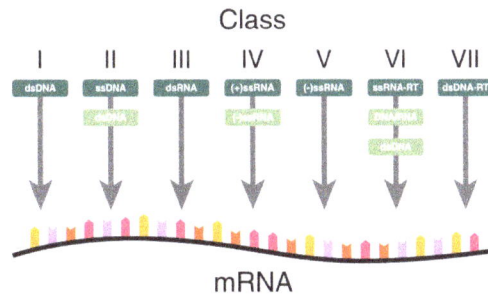

The Baltimore Classification of viruses is based on the method of viral mRNA synthesis

The Nobel Prize-winning biologist David Baltimore devised the Baltimore classification system. The ICTV classification system is used in conjunction with the Baltimore classification system in modern virus classification.

The Baltimore classification of viruses is based on the mechanism of mRNA production. Viruses must generate mRNAs from their genomes to produce proteins and replicate themselves, but different mechanisms are used to achieve this in each virus family. Viral genomes may be single-stranded (ss) or double-stranded (ds), RNA or DNA, and may or may not use reverse transcriptase (RT). In addition, ssRNA viruses may be either sense (+) or antisense (−). This classification places viruses into seven groups:

- I: dsDNA viruses (e.g. Adenoviruses, Herpesviruses, Poxviruses)

- II: ssDNA viruses (+ strand or "sense") DNA (e.g. Parvoviruses)

- III: dsRNA viruses (e.g. Reoviruses)

- IV: (+)ssRNA viruses (+ strand or sense) RNA (e.g. Picornaviruses, Togaviruses)

- V: (−)ssRNA viruses (− strand or antisense) RNA (e.g. Orthomyxoviruses, Rhabdoviruses)

- VI: ssRNA-RT viruses (+ strand or sense) RNA with DNA intermediate in life-cycle (e.g. Retroviruses)

- VII: dsDNA-RT viruses (e.g. Hepadnaviruses)

As an example of viral classification, the chicken pox virus, varicella zoster (VZV), belongs to the order *Herpesvirales*, family *Herpesviridae*, subfamily *Alphaherpesvirinae*, and genus *Varicellovirus*. VZV is in Group I of the Baltimore Classification because it is a dsDNA virus that does not use reverse transcriptase.

The complete set of viruses in an organism or habitat is called the virome; for example, all human viruses constitute the human virome.

Role in Human Disease

Examples of common human diseases caused by viruses include the common cold, influenza, chickenpox, and cold sores. Many serious diseases such as Ebola virus disease, AIDS, avian influenza, and SARS are caused by viruses. The relative ability of viruses to cause disease is described in terms of virulence. Other diseases are under investigation to discover if they have a virus as the causative agent, such as the possible connection between human herpesvirus 6 (HHV6) and neurological diseases such as multiple sclerosis and chronic fatigue syndrome. There is controversy over whether the bornavirus, previously thought to cause neurological diseases in horses, could be responsible for psychiatric illnesses in humans.

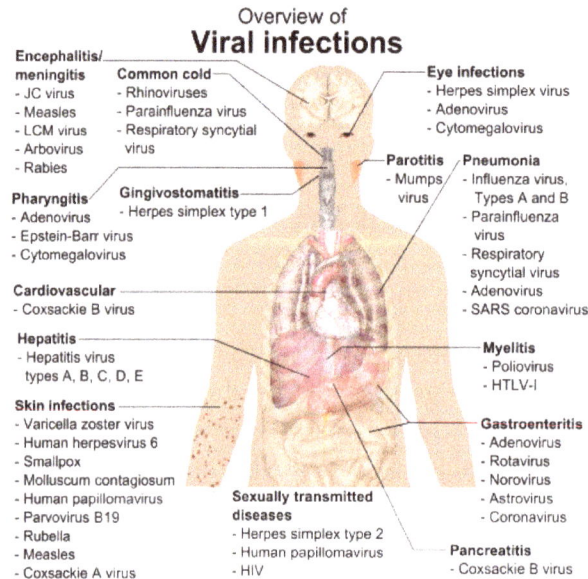

Overview of
Viral infections

Encephalitis/
meningitis
- JC virus
- Measles
- LCM virus
- Arbovirus
- Rabies

Common cold
- Rhinoviruses
- Parainfluenza virus
- Respiratory syncytial virus

Eye infections
- Herpes simplex virus
- Adenovirus
- Cytomegalovirus

Parotitis
- Mumps virus

Pneumonia
- Influenza virus, Types A and B
- Parainfluenza virus
- Respiratory syncytial virus
- Adenovirus
- SARS coronavirus

Pharyngitis
- Adenovirus
- Epstein-Barr virus
- Cytomegalovirus

Gingivostomatitis
- Herpes simplex type 1

Cardiovascular
- Coxsackie B virus

Hepatitis
- Hepatitis virus types A, B, C, D, E

Skin infections
- Varicella zoster virus
- Human herpesvirus 6
- Smallpox
- Molluscum contagiosum
- Human papillomavirus
- Parvovirus B19
- Rubella
- Measles
- Coxsackie A virus

Myelitis
- Poliovirus
- HTLV-I

Gastroenteritis
- Adenovirus
- Rotavirus
- Norovirus
- Astrovirus
- Coronavirus

Sexually transmitted diseases
- Herpes simplex type 2
- Human papillomavirus
- HIV

Pancreatitis
- Coxsackie B virus

Overview of the main types of viral infection and the most notable species involved

Viruses have different mechanisms by which they produce disease in an organism, which depends largely on the viral species. Mechanisms at the cellular level primarily include cell lysis, the breaking open and subsequent death of the cell. In multicellular organisms, if enough cells die, the whole organism will start to suffer the effects. Although viruses cause disruption of healthy homeostasis, resulting in disease, they may exist relatively harmlessly within an organism. An example would include the ability of the herpes simplex virus, which causes cold sores, to remain in a dormant state within the human body. This is called latency and is a characteristic of the herpes viruses, including Epstein–Barr virus, which causes glandular fever, and varicella zoster virus, which causes chickenpox and shingles. Most people have been infected with at least one of these types of herpes virus. However, these latent viruses might sometimes be beneficial, as the presence of the virus can increase immunity against bacterial pathogens, such as *Yersinia pestis*.

Some viruses can cause lifelong or chronic infections, where the viruses continue to replicate in the body despite the host's defence mechanisms. This is common in hepatitis B virus and hepatitis C virus infections. People chronically infected are known as carriers, as they serve as reservoirs of infectious virus. In populations with a high proportion of carriers, the disease is said to be endemic.

Epidemiology

Viral epidemiology is the branch of medical science that deals with the transmission and control of virus infections in humans. Transmission of viruses can be vertical, which means from mother to child, or horizontal, which means from person to person. Examples of vertical transmission include hepatitis B virus and HIV, where the baby is born already infected with the virus. Another, more rare, example is the varicella zoster virus, which, although causing relatively mild infections in humans, can be fatal to the foetus and newborn baby.

Horizontal transmission is the most common mechanism of spread of viruses in populations. Transmission can occur when: body fluids are exchanged during sexual activity, e.g., HIV; blood is exchanged by contaminated transfusion or needle sharing, e.g., hepatitis C; exchange of saliva by mouth, e.g., Epstein–Barr virus; contaminated food or water is ingested, e.g., norovirus; aerosols containing virions are inhaled, e.g., influenza virus; and insect vectors such as mosquitoes penetrate the skin of a host, e.g., dengue. The rate or speed of transmission of viral infections depends on factors that include population density, the number of susceptible individuals, (i.e., those not immune), the quality of healthcare and the weather.

Epidemiology is used to break the chain of infection in populations during outbreaks of viral diseases. Control measures are used that are based on knowledge of how the virus is transmitted. It is important to find the source, or sources, of the outbreak and to identify the virus. Once the virus has been identified, the chain of transmission can sometimes be broken by vaccines. When vaccines are not available, sanitation and disinfection can be effective. Often, infected people are isolated from the rest of the community, and those that have been exposed to the virus are placed in quarantine. To control the outbreak of foot-and-mouth disease in cattle in Britain in 2001, thousands of cattle were slaughtered. Most viral infections of humans and other animals have incubation periods during which the infection causes no signs or symptoms. Incubation periods for viral diseases range from a few days to weeks, but are known for most infections. Somewhat overlapping, but mainly following the incubation period, there is a period of communicability — a time when an infected individual or animal is contagious and can infect another person or animal. This, too, is known for many viral infections, and knowledge of the length of both periods is important in the control of outbreaks. When outbreaks cause an unusually high proportion of cases in a population, community, or region, they are called epidemics. If outbreaks spread worldwide, they are called pandemics.

Epidemics and Pandemics

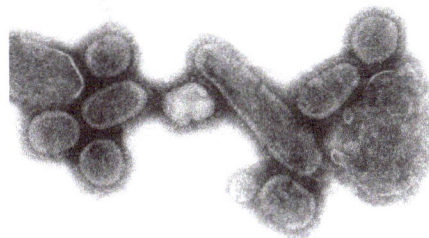

Transmission electron microscope image of a recreated 1918 influenza virus

Native American populations were devastated by contagious diseases, in particular, smallpox, brought to the Americas by European colonists. It is unclear how many Native Americans were killed by foreign diseases after the arrival of Columbus in the Americas, but the numbers have been estimated to be close to 70% of the indigenous population. The damage done by this disease significantly aided European attempts to displace and conquer the native population.

A pandemic is a worldwide epidemic. The 1918 flu pandemic, which lasted until 1919, was a category 5 influenza pandemic caused by an unusually severe and deadly influenza A virus. The victims were often healthy young adults, in contrast to most influenza outbreaks, which predominantly affect juvenile, elderly, or otherwise-weakened patients. Older estimates say it killed 40–50 million people, while more recent research suggests that it may have killed as many as 100 million people, or 5% of the world's population in 1918.

Most researchers believe that HIV originated in sub-Saharan Africa during the 20th century; it is now a pandemic, with an estimated 38.6 million people now living with the disease worldwide. The Joint United Nations Programme on HIV/AIDS (UNAIDS) and the World Health Organization (WHO) estimate that AIDS has killed more than 25 million people since it was first recognised on 5 June 1981, making it one of the most destructive epidemics in recorded history. In 2007 there were 2.7 million new HIV infections and 2 million HIV-related deaths.

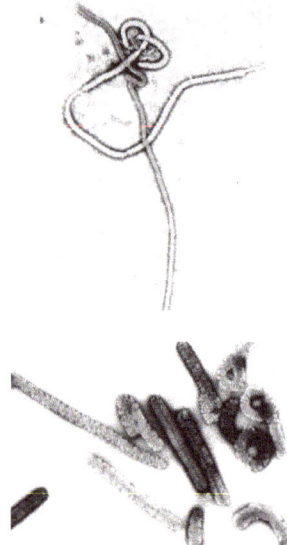

Ebola (top) and Marburg viruses (bottom)

Several highly lethal viral pathogens are members of the *Filoviridae*. Filoviruses are filament-like viruses that cause viral hemorrhagic fever, and include ebolaviruses and marburgviruses. Marburg virus, first discovered in 1967, attracted widespread press attention in April 2005 for an outbreak in Angola. Ebola Virus Disease has also caused intermittent outbreaks with high mortality rates since 1976 when it was first identified. The worst and most recent one is the West Africa epidemic.

Cancer

Viruses are an established cause of cancer in humans and other species. Viral cancers occur only in a minority of infected persons (or animals). Cancer viruses come from a range of virus families, including both RNA and DNA viruses, and so there is no single type of "oncovirus" (an obsolete term originally used for acutely transforming retroviruses). The development of cancer is determined by a variety of factors such as host immunity and mutations in the host. Viruses accepted to cause human cancers include some genotypes of human papillomavirus, hepatitis B virus, hepatitis C virus, Epstein–Barr virus, Kaposi's sarcoma-associated herpesvirus and human T-lymphotropic virus. The most recently discovered human cancer virus is a polyomavirus (Merkel cell polyoma-virus) that causes most cases of a rare form of skin cancer called Merkel cell carcinoma. Hepatitis viruses can develop into a chronic viral infection that leads to liver cancer. Infection by human T-lymphotropic virus can lead to tropical spastic paraparesis and adult T-cell leukaemia. Human papillomaviruses are an established cause of cancers of cervix, skin, anus, and penis. Within the *Herpesviridae*, Kaposi's sarcoma-associated herpesvirus causes Kaposi's sarcoma and body-cavity lymphoma, and Epstein–Barr virus causes Burkitt's lymphoma, Hodgkin's lymphoma, B lymph-oproliferative disorder, and nasopharyngeal carcinoma. Merkel cell polyomavirus closely related to SV40 and mouse polyomaviruses that have been used as animal models for cancer viruses for over 50 years.

Host Defence Mechanisms

The body's first line of defence against viruses is the innate immune system. This comprises cells and other mechanisms that defend the host from infection in a non-specific manner. This means that the cells of the innate system recognise, and respond to, pathogens in a generic way, but, unlike the adaptive immune system, it does not confer long-lasting or protective immunity to the host.

RNA interference is an important innate defence against viruses. Many viruses have a replication strategy that involves double-stranded RNA (dsRNA). When such a virus infects a cell, it releases its RNA molecule or molecules, which immediately bind to a protein complex called a dicer that cuts the RNA into smaller pieces. A biochemical pathway – the RISC complex, is activated, which ensures cell survival by degrading the viral mRNA. Rotaviruses have evolved to avoid this defence mechanism by not uncoating fully inside the cell, and releasing newly produced mRNA through pores in the particle's inner capsid. Their genomic dsRNA remains protected inside the core of the virion.

When the adaptive immune system of a vertebrate encounters a virus, it produces specific antibod-ies that bind to the virus and often render it non-infectious. This is called humoral immunity. Two types of antibodies are important. The first, called IgM, is highly effective at neutralising viruses but is produced by the cells of the immune system only for a few weeks. The second, called IgG, is produced indefinitely. The presence of IgM in the blood of the host is used to test for acute in-fection, whereas IgG indicates an infection sometime in the past. IgG antibody is measured when tests for immunity are carried out.

Antibodies can continue to be an effective defence mechanism even after viruses have managed to gain entry to the host cell. A protein that is in cells, called TRIM21, can attach to the antibodies on the surface of the virus particle. This primes the subsequent destruction of the virus by the en-zymes of the cell's proteosome system.

A second defence of vertebrates against viruses is called cell-mediated immunity and involves im-mune cells known as T cells. The body's cells constantly display short fragments of their proteins on the cell's surface, and, if a T cell recognises a suspicious viral fragment there, the host cell is destroyed by *killer T* cells and the virus-specific T-cells proliferate. Cells such as the macrophage are specialists at this antigen presentation. The production of interferon is an important host de-fence mechanism. This is a hormone produced by the body when viruses are present. Its role in immunity is complex; it eventually stops the viruses from reproducing by killing the infected cell and its close neighbours.

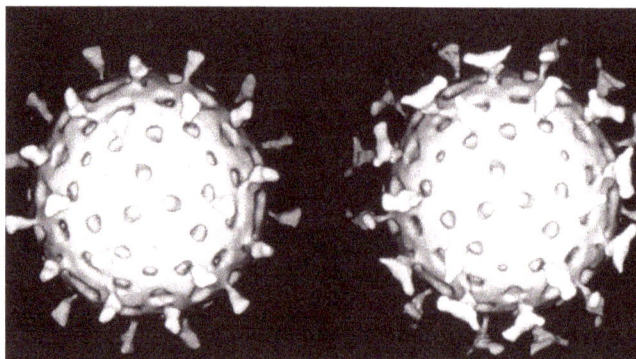

Two rotaviruses: the one on the right is coated with antibodies that stop its attaching to cells and infecting them

Not all virus infections produce a protective immune response in this way. HIV evades the immune system by constantly changing the amino acid sequence of the proteins on the surface of the virion. This is known as "escape mutation" as the viral epitopes escape recognition by the host immune response. These persistent viruses evade immune control by sequestration, blockade of antigen presentation, cytokine resistance, evasion of natural killer cell activities, escape from apoptosis, and antigenic shift. Other viruses, called *neurotropic viruses*, are disseminated by neural spread where the immune system may be unable to reach them.

Prevention and Treatment

Because viruses use vital metabolic pathways within host cells to replicate, they are difficult to eliminate without using drugs that cause toxic effects to host cells in general. The most effective medical approaches to viral diseases are vaccinations to provide immunity to infection, and anti-viral drugs that selectively interfere with viral replication.

Vaccines

Vaccination is a cheap and effective way of preventing infections by viruses. Vaccines were used to prevent viral infections long before the discovery of the actual viruses. Their use has resulted in a dramatic decline in morbidity (illness) and mortality (death) associated with viral infections such as polio, measles, mumps and rubella. Smallpox infections have been eradicated. Vaccines are available to prevent over thirteen viral infections of humans, and more are used to prevent viral infections of animals. Vaccines can consist of live-attenuated or killed viruses, or viral proteins (antigens). Live vaccines contain weakened forms of the virus, which do not cause the disease but, nonetheless, confer immunity. Such viruses are called attenuated. Live vaccines can be dangerous when given to people with a weak immunity (who are described as immunocompromised), because in these people, the weakened virus can cause the original disease. Biotechnology and genetic engineering techniques are used to produce subunit vaccines. These vaccines use only the capsid proteins of the virus. Hepatitis B vaccine is an example of this type of vaccine. Subunit vaccines are safe for immunocompromised patients because they cannot cause the disease. The yellow fever virus vaccine, a live-attenuated strain called 17D, is probably the safest and most effective vaccine ever generated.

Antiviral Drugs

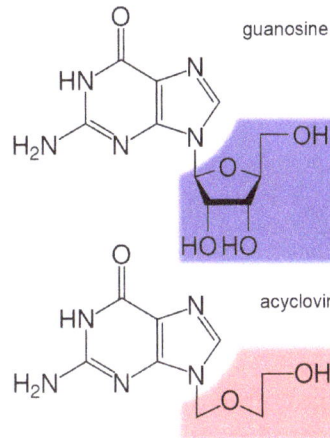

The structure of the DNA base guanosine and the antiviral drug acyclovir

Antiviral drugs are often nucleoside analogues (fake DNA building-blocks), which viruses mistakenly incorporate into their genomes during replication. The life-cycle of the virus is then halted because the newly synthesised DNA is inactive. This is because these analogues lack the hydroxyl groups, which, along with phosphorus atoms, link together to form the strong "backbone" of the DNA molecule. This is called DNA chain termination. Examples of nucleoside analogues are aciclovir for Herpes simplex virus infections and lamivudine for HIV and Hepatitis B virus infections. Aciclovir is one of the oldest and most frequently prescribed antiviral drugs. Other antiviral drugs in use target different stages of the viral life cycle. HIV is dependent on a proteolytic enzyme called the HIV-1 protease for it to become fully infectious. There is a large class of drugs called protease inhibitors that inactivate this enzyme.

Hepatitis C is caused by an RNA virus. In 80% of people infected, the disease is chronic, and without treatment, they are infected for the remainder of their lives. However, there is now an effective treatment that uses the nucleoside analogue drug ribavirin combined with interferon. The treatment of chronic carriers of the hepatitis B virus by using a similar strategy using lamivudine has been developed.

Infection in Other Species

Viruses infect all cellular life and, although viruses occur universally, each cellular species has its own specific range that often infect only that species. Some viruses, called satellites, can replicate only within cells that have already been infected by another virus.

Animal Viruses

Viruses are important pathogens of livestock. Diseases such as foot-and-mouth disease and bluetongue are caused by viruses. Companion animals such as cats, dogs, and horses, if not vaccinated, are susceptible to serious viral infections. Canine parvovirus is caused by a small DNA virus and infections are often fatal in pups. Like all invertebrates, the honey bee is susceptible to many viral infections. However, most viruses co-exist harmlessly in their host and cause no signs or symptoms of disease.

Plant Viruses

Peppers infected by mild mottle virus

There are many types of plant virus, but often they cause only a loss of yield, and it is not economically viable to try to control them. Plant viruses are often spread from plant to plant by organisms, known as *vectors*. These are normally insects, but some fungi, nematode worms, and single-celled organisms have been shown to be vectors. When control of plant virus infections is considered economical, for perennial fruits, for example, efforts are concentrated on killing the vectors and removing alternate hosts such as weeds. Plant viruses cannot infect humans and other animals because they can reproduce only in living plant cells.

Plants have elaborate and effective defence mechanisms against viruses. One of the most effective is the presence of so-called resistance (R) genes. Each R gene confers resistance to a particular virus by triggering localised areas of cell death around the infected cell, which can often be seen with the unaided eye as large spots. This stops the infection from spreading. RNA interference is also an effective defence in plants. When they are infected, plants often produce natural disinfectants that kill viruses, such as salicylic acid, nitric oxide, and reactive oxygen molecules.

Plant virus particles or virus-like particles (VLPs) have applications in both biotechnology and nanotechnology. The capsids of most plant viruses are simple and robust structures and can be produced in large quantities either by the infection of plants or by expression in a variety of heterologous systems. Plant virus particles can be modified genetically and chemically to encapsulate foreign material and can be incorporated into supramolecular structures for use in biotechnology.

Bacterial Viruses

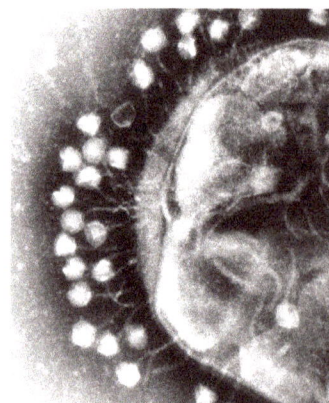

Transmission electron micrograph of multiple bacteriophages attached to a bacterial cell wall

Bacteriophages are a common and diverse group of viruses and are the most abundant form of biological entity in aquatic environments – there are up to ten times more of these viruses in the oceans than there are bacteria, reaching levels of 250,000,000 bacteriophages per millilitre of seawater. These viruses infect specific bacteria by binding to surface receptor molecules and then entering the cell. Within a short amount of time, in some cases just minutes, bacterial polymerase starts translating viral mRNA into protein. These proteins go on to become either new virions within the cell, helper proteins, which help assembly of new virions, or proteins involved in cell lysis. Viral enzymes aid in the breakdown of the cell membrane, and, in the case of the T4 phage, in just over twenty minutes after injection over three hundred phages could be released.

The major way bacteria defend themselves from bacteriophages is by producing enzymes that destroy foreign DNA. These enzymes, called restriction endonucleases, cut up the viral DNA that bacteriophages inject into bacterial cells. Bacteria also contain a system that uses CRISPR sequences to retain fragments of the genomes of viruses that the bacteria have come into contact with in the past, which allows them to block the virus's replication through a form of RNA interference. This genetic system provides bacteria with acquired immunity to infection.

Archaean Viruses

Some viruses replicate within archaea: these are double-stranded DNA viruses with unusual and sometimes unique shapes. These viruses have been studied in most detail in the thermophilic archaea, particularly the orders Sulfolobales and Thermoproteales. Defences against these viruses involve RNA interference from repetitive DNA sequences within archaean genomes that are related to the genes of the viruses. Most archaea have CRISPR–Cas systems as an adaptive defence against viruses. These enable archaea to retain sections of viral DNA, which are then used to target and eliminate subsequent infections by the virus using a process similar to RNA interference.

Role in Aquatic Ecosystems

A teaspoon of seawater contains about one million viruses. Most of these are bacteriophages, which are harmless to plants and animals, and are in fact essential to the regulation of saltwater and freshwater ecosystems. They infect and destroy bacteria in aquatic microbial communities, and are the most important mechanism of recycling carbon in the marine environment. The organic molecules released from the dead bacterial cells stimulate fresh bacterial and algal growth. Viral activity may also contribute to the biological pump, the process whereby carbon is sequestered in the deep ocean.

Microorganisms constitute more than 90% of the biomass in the sea. It is estimated that viruses kill approximately 20% of this biomass each day and that there are 15 times as many viruses in the oceans as there are bacteria and archaea. Viruses are the main agents responsible for the rapid destruction of harmful algal blooms, which often kill other marine life. The number of viruses in the oceans decreases further offshore and deeper into the water, where there are fewer host organisms.

Like any organism, marine mammals are susceptible to viral infections. In 1988 and 2002, thousands of harbour seals were killed in Europe by phocine distemper virus. Many other viruses, including caliciviruses, herpesviruses, adenoviruses and parvoviruses, circulate in marine mammal populations.

Role in Evolution

Viruses are an important natural means of transferring genes between different species, which increases genetic diversity and drives evolution. It is thought that viruses played a central role in the early evolution, before the diversification of bacteria, archaea and eukaryotes, at the time of the last universal common ancestor of life on Earth. Viruses are still one of the largest reservoirs of unexplored genetic diversity on Earth.

Applications

Life Sciences and Medicine

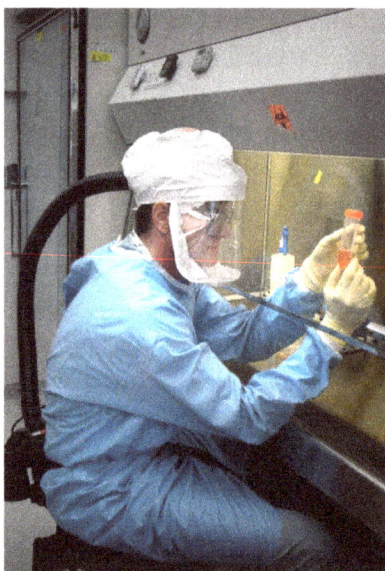

Scientist studying the H5N1 influenza virus

Viruses are important to the study of molecular and cell biology as they provide simple systems that can be used to manipulate and investigate the functions of cells. The study and use of viruses have provided valuable information about aspects of cell biology. For example, viruses have been useful in the study of genetics and helped our understanding of the basic mechanisms of molecular genetics, such as DNA replication, transcription, RNA processing, translation, protein transport, and immunology.

Geneticists often use viruses as vectors to introduce genes into cells that they are studying. This is useful for making the cell produce a foreign substance, or to study the effect of introducing a new gene into the genome. In similar fashion, virotherapy uses viruses as vectors to treat various diseases, as they can specifically target cells and DNA. It shows promising use in the treatment of cancer and in gene therapy. Eastern European scientists have used phage therapy as an alternative to antibiotics for some time, and interest in this approach is increasing, because of the high level of antibiotic resistance now found in some pathogenic bacteria. Expression of heterologous proteins by viruses is the basis of several manufacturing processes that are currently being used for the production of various proteins such as vaccine antigens and antibodies. Industrial processes have been recently developed using viral vectors and a number of pharmaceutical proteins are currently in pre-clinical and clinical trials.

Virotherapy

Virotherapy involves the use of genetically modified viruses to treat diseases. Viruses have been modified by scientists to reproduce in cancer cells and destroy them but not infect healthy cells. Talimogene laherparepvec (T-VEC), for example, is a modified herpes simplex virus that has had a gene, which is required for viruses to replicate in healthy cells, deleted and replaced with a human gene (GM-CSF) that stimulates immunity. When this virus infects cancer cells, it destroys them and in doing so the presence the GM-CSF gene attracts dendritic cells from the surrounding tissues of the body. The dendritic cells process the dead cancer cells and present components of them to other cells of the immune system. Having completed successful clinical trials, this virus is expected to gain approval for the treatment of a skin cancer called melanoma in late 2015. Viruses that have been reprogrammed to kill cancer cells are called oncolytic viruses.

Materials Science and Nanotechnology

Current trends in nanotechnology promise to make much more versatile use of viruses. From the viewpoint of a materials scientist, viruses can be regarded as organic nanoparticles. Their surface carries specific tools designed to cross the barriers of their host cells. The size and shape of viruses, and the number and nature of the functional groups on their surface, is precisely defined. As such, viruses are commonly used in materials science as scaffolds for covalently linked surface modifications. A particular quality of viruses is that they can be tailored by directed evolution. The powerful techniques developed by life sciences are becoming the basis of engineering approaches towards nanomaterials, opening a wide range of applications far beyond biology and medicine.

Because of their size, shape, and well-defined chemical structures, viruses have been used as templates for organising materials on the nanoscale. Recent examples include work at the Naval Research Laboratory in Washington, D.C., using Cowpea mosaic virus (CPMV) particles to amplify signals in DNA microarray based sensors. In this application, the virus particles separate the fluorescent dyes used for signalling to prevent the formation of non-fluorescent dimers that act as quenchers. Another example is the use of CPMV as a nanoscale breadboard for molecular electronics.

Synthetic Viruses

Many viruses can be synthesised de novo ("from scratch") and the first synthetic virus was created in 2002. Although somewhat of a misconception, it is not the actual virus that is synthesised, but rather its DNA genome (in case of a DNA virus), or a cDNA copy of its genome (in case of RNA viruses). For many virus families the naked synthetic DNA or RNA (once enzymatically converted back from the synthetic cDNA) is infectious when introduced into a cell. That is, they contain all the necessary information to produce new viruses. This technology is now being used to investigate novel vaccine strategies. The ability to synthesise viruses has far-reaching consequences, since viruses can no longer be regarded as extinct, as long as the information of their genome sequence is known and permissive cells are available. As of March 2014, the full-length genome sequences of 3843 different viruses, including smallpox, are publicly available in an online database maintained by the National Institutes of Health.

Weapons

The ability of viruses to cause devastating epidemics in human societies has led to the concern that viruses could be weaponised for biological warfare. Further concern was raised by the successful recreation of the infamous 1918 influenza virus in a laboratory. Smallpox virus devastated numerous societies throughout history before its eradication. There are only two centres in the world that are authorised by the WHO to keep stocks of smallpox virus: the Vector Institute in Russia and the Centers for Disease Control and Prevention in the United States. Fears that it may be used as a weapon may not be totally unfounded. As the vaccine for smallpox sometimes had severe side-effects, it is no longer used routinely in any country. Thus, much of the modern human population has almost no established resistance to smallpox, and would be vulnerable to the virus.

Bacteria

Bacteria (common noun bacteria, singular bacterium) constitute a large domain of prokaryotic microorganisms. Typically a few micrometres in length, bacteria have a number of shapes, ranging from spheres to rods and spirals. Bacteria were among the first life forms to appear on Earth, and are present in most of its habitats. Bacteria inhabit soil, water, acidic hot springs, radioactive waste, and the deep portions of Earth's crust. Bacteria also live in symbiotic and parasitic relationships with plants and animals.

There are typically 40 million bacterial cells in a gram of soil and a million bacterial cells in a millilitre of fresh water. There are approximately 5×10^{30} bacteria on Earth, forming a biomass which exceeds that of all plants and animals. Bacteria are vital in recycling nutrients, with many of the stages in nutrient cycles dependent on these organisms, such as the fixation of nitrogen from the atmosphere and putrefaction. In the biological communities surrounding hydrothermal vents and cold seeps, bacteria provide the nutrients needed to sustain life by converting dissolved compounds, such as hydrogen sulphide and methane, to energy. On 17 March 2013, researchers reported data that suggested bacterial life forms thrive in the Mariana Trench, which with a depth of up to 11 kilometres is the deepest part of the Earth's oceans. Other researchers reported related studies that microbes thrive inside rocks up to 580 metres below the sea floor under 2.6 kilometres of ocean off the coast of the northwestern United States. According to one of the researchers, "You can find microbes everywhere — they're extremely adaptable to conditions, and survive wherever they are."

Most bacteria have not been characterised, and only about half of the bacterial phyla have species that can be grown in the laboratory. The study of bacteria is known as bacteriology, a branch of microbiology.

There are approximately ten times as many bacterial cells in the human flora as there are human cells in the body, with the largest number of the human flora being in the gut flora, and a large number on the skin. The vast majority of the bacteria in the body are rendered harmless by the protective effects of the immune system, and some are beneficial. However, several species of bacteria are pathogenic and cause infectious diseases, including cholera, syphilis, anthrax, leprosy, and bubonic plague. The most common fatal bacterial diseases are respiratory infections, with tuberculosis alone killing about 2 million people per year, mostly in sub-Saharan Africa. In developed countries, antibiotics are used to treat bacterial infections and are also used in farming, making antibiotic resistance a growing problem. In industry, bacteria are important in sewage treatment and the breakdown of oil spills, the production of cheese and yogurt through fermentation, and the recovery of gold, palladium, copper and other metals in the mining sector, as well as in biotechnology, and the manufacture of antibiotics and other chemicals.

Once regarded as plants constituting the class *Schizomycetes*, bacteria are now classified as prokaryotes. Unlike cells of animals and other eukaryotes, bacterial cells do not contain a nucleus and rarely harbour membrane-bound organelles. Although the term *bacteria* traditionally included all prokaryotes, the scientific classification changed after the discovery in the 1990s that prokaryotes consist of two very different groups of organisms that evolved from an ancient common ancestor. These evolutionary domains are called *Bacteria* and *Archaea*.

Origin and Early Evolution

The ancestors of modern bacteria were unicellular microorganisms that were the first forms of life to appear on Earth, about 4 billion years ago. For about 3 billion years, most organisms were microscopic, and bacteria and archaea were the dominant forms of life. In 2008, fossils of macroorganisms were discovered and named as the Francevillian biota. Although bacterial fossils exist, such as stromatolites, their lack of distinctive morphology prevents them from being used to examine the history of bacterial evolution, or to date the time of origin of a particular bacterial species. However, gene sequences can be used to reconstruct the bacterial phylogeny, and these studies indicate that bacteria diverged first from the archaeal/eukaryotic lineage. Bacteria were also involved in the second great evolutionary divergence, that of the archaea and eukaryotes. Here, eukaryotes resulted from the entering of ancient bacteria into endosymbiotic associations with the ancestors of eukaryotic cells, which were themselves possibly related to the Archaea. This involved the engulfment by proto-eukaryotic cells of alphaproteobacterial symbionts to form either mitochondria or hydrogenosomes, which are still found in all known Eukarya (sometimes in highly reduced form, e.g. in ancient "amitochondrial" protozoa). Later on, some eukaryotes that already contained mitochondria also engulfed cyanobacterial-like organisms. This led to the formation of chloroplasts in algae and plants. There are also some algae that originated from even later endosymbiotic events. Here, eukaryotes engulfed a eukaryotic algae that developed into a "second-generation" plastid. This is known as secondary endosymbiosis.

Morphology

Bacteria display a wide diversity of shapes and sizes, called *morphologies*. Bacterial cells are about one-tenth the size of eukaryotic cells and are typically 0.5–5.0 micrometres in length. However, a few species are visible to the unaided eye — for example, *Thiomargarita namibiensis* is up to half a millimetre long and *Epulopiscium fishelsoni* reaches 0.7 mm. Among the smallest bacteria are members of the genus *Mycoplasma*, which measure only 0.3 micrometres, as small as the largest viruses. Some bacteria may be even smaller, but these ultramicrobacteria are not well-studied.

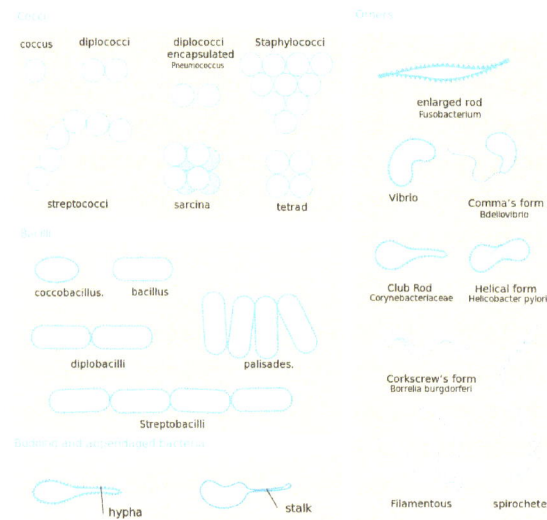

Bacteria display many cell morphologies and arrangements

Most bacterial species are either spherical, called *cocci*, or rod-shaped, called *bacilli*. Elongation is associated with swimming. Some bacteria, called *vibrio*, are shaped like slightly curved rods or comma-shaped; others can be spiral-shaped, called *spirilla*, or tightly coiled, called *spirochaetes*. A small number of species even have tetrahedral or cuboidal shapes. More recently, some bacteria were discovered deep under Earth's crust that grow as branching filamentous types with a star-shaped cross-section. The large surface area to volume ratio of this morphology may give these bacteria an advantage in nutrient-poor environments. This wide variety of shapes is determined by the bacterial cell wall and cytoskeleton, and is important because it can influence the ability of bacteria to acquire nutrients, attach to surfaces, swim through liquids and escape predators.

A biofilm of thermophilic bacteria in the outflow of Mickey Hot Springs, Oregon, approximately 20 mm thick.

Many bacterial species exist simply as single cells, others associate in characteristic patterns: *Neisseria* form diploids (pairs), *Streptococcus* form chains, and *Staphylococcus* group together in "bunch of grapes" clusters. Bacteria can also be elongated to form filaments, for example the Actinobacteria. Filamentous bacteria are often surrounded by a sheath that contains many individual cells. Certain types, such as species of the genus *Nocardia*, even form complex, branched filaments, similar in appearance to fungal mycelia.

Bacteria often attach to surfaces and form dense aggregations called *biofilms* or bacterial mats. These films can range from a few micrometres in thickness to up to half a metre in depth, and may contain multiple species of bacteria, protists and archaea. Bacteria living in biofilms display a complex arrangement of cells and extracellular components, forming secondary structures, such as microcolonies, through which there are networks of channels to enable better diffusion of nutrients. In natural environments, such as soil or the surfaces of plants, the majority of bacteria are bound to surfaces in biofilms. Biofilms are also important in medicine, as these structures are often present during chronic bacterial infections or in infections of implanted medical devices, and bacteria protected within biofilms are much harder to kill than individual isolated bacteria.

Even more complex morphological changes are sometimes possible. For example, when starved of amino acids, Myxobacteria detect surrounding cells in a process known as quorum sensing, migrate towards each other, and aggregate to form fruiting bodies up to 500 micrometres long and containing approximately 100,000 bacterial cells. In these fruiting bodies, the bacteria perform separate tasks; this type of cooperation is a simple type of multicellular organisation. For example, about one in 10 cells migrate to the top of these fruiting bodies and differentiate into a specialised dormant state called myxospores, which are more resistant to drying and other adverse environmental conditions than are ordinary cells.

Cellular Structure

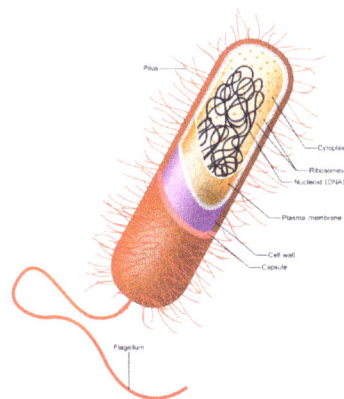

Structure and contents of a typical gram-positive bacterial cell (seen by the fact that only *one* cell membrane is present).

Intracellular Structures

The bacterial cell is surrounded by a cell membrane (also known as a lipid, cytoplasmic or plasma membrane). This membrane encloses the contents of the cell and acts as a barrier to hold nutrients, proteins and other essential components of the *cytoplasm* within the cell. As they are prokaryotes, bacteria do not usually have membrane-bound organelles in their cytoplasm, and thus contain few large intracellular structures. They lack a true nucleus, mitochondria, chloroplasts and the other organelles present in eukaryotic cells. Bacteria were once seen as simple bags of cytoplasm, but structures such as the *prokaryotic cytoskeleton* and the localisation of proteins to specific locations within the cytoplasm that give bacteria some complexity have been discovered. These subcellular levels of organisation have been called "bacterial hyperstructures".

Bacterial microcompartments, such as carboxysomes, provide a further level of organisation; they are compartments within bacteria that are surrounded by polyhedral protein shells, rather than by lipid membranes. These "polyhedral organelles" localise and compartmentalise bacterial metabolism, a function performed by the membrane-bound organelles in eukaryotes.

Many important biochemical reactions, such as energy generation, use concentration gradients across membranes. The general lack of internal membranes in bacteria means reactions such as electron transport occur across the cell membrane between the cytoplasm and the *periplasmic space*. However, in many photosynthetic bacteria the plasma membrane is highly folded and fills most of the cell with layers of light-gathering membrane. These light-gathering complexes may even form lipid-enclosed structures called chlorosomes in green sulfur bacteria. Other proteins import nutrients across the cell membrane, or expel undesired molecules from the cytoplasm.

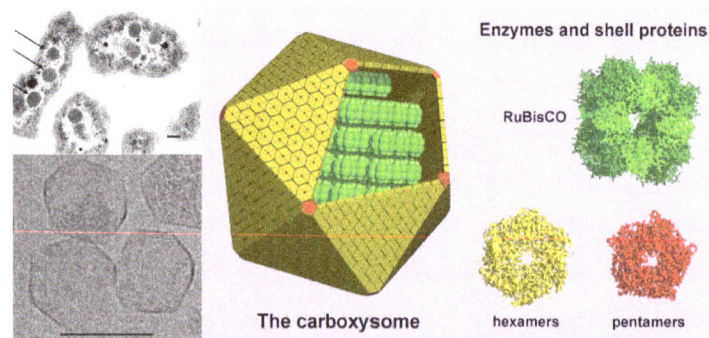

Carboxysomes are protein-enclosed bacterial organelles. Top left is an electron microscope image of carboxysomes in *Halothiobacillus neapolitanus*, below is an image of purified carboxysomes. On the right is a model of their structure. Scale bars are 100 nm.

Bacteria do not have a membrane-bound nucleus, and their genetic material is typically a single circular DNA chromosome located in the cytoplasm in an irregularly shaped body called the *nucleoid*. The nucleoid contains the chromosome with its associated proteins and RNA. The phylum Planctomycetes and candidate phylum Poribacteria may be exceptions to the general absence of internal membranes in bacteria, because they appear to have a double membrane around their nucleoids and contain other membrane-bound cellular structures. Like all living organisms, bacteria contain *ribosomes*, often grouped in chains called polyribosomes, for the production of proteins, but the structure of the bacterial ribosome is different from that of eukaryotes and Archaea. Bacterial ribosomes have a sedimentation rate of 70S (measured in Svedberg units): their subunits have rates of 30S and 50S. Some antibiotics bind specifically to 70S ribosomes and inhibit bacterial protein synthesis. Those antibiotics kill bacteria without affecting the larger 80S ribosomes of eukaryotic cells and without harming the host.

Some bacteria produce intracellular nutrient storage granules for later use, such as glycogen, polyphosphate, sulfur or polyhydroxyalkanoates. Certain bacterial species, such as the photosynthetic Cyanobacteria, produce internal gas vesicles, which they use to regulate their buoyancy – allowing them to move up or down into water layers with different light intensities and nutrient levels. *Intracellular membranes* called *chromatophores* are also found in membranes of phototrophic bacteria. Used primarily for photosynthesis, they contain bacteriochlorophyll pigments and carotenoids. An early idea was that bacteria might contain membrane folds termed mesosomes, but these were later shown to be artefacts produced by the chemicals used to prepare the cells for electron microscopy. *Inclusions* are considered to be nonliving components of the cell that do not possess metabolic activity and are not bounded by membranes. The most common inclusions are glycogen, lipid droplets, crystals, and pigments. *Volutin granules* are cytoplasmic inclusions of complexed inorganic polyphosphate. These granules are called *metachromatic granules* due to their displaying the metachromatic effect; they appear red or blue when stained with the blue dyes methylene blue or toluidine blue. *Gas vacuoles*, which are freely permeable to gas, are membrane-bound vesicles present in some species of *Cyanobacteria*. They allow the bacteria to control their buoyancy. *Microcompartments* are widespread, membrane-bound organelles that are made of a protein shell that surrounds and encloses various enzymes. *Carboxysomes* are bacterial microcompartments that contain enzymes involved in carbon fixation. *Magnetosomes* are bacterial microcompartments, present in magnetotactic bacteria, that contain magnetic crystals.

Extracellular Structures

In most bacteria, a *cell wall* is present on the outside of the cell membrane. The cell membrane and cell wall comprise the *cell envelope*. A common bacterial cell wall material is *peptidoglycan* (called "murein" in older sources), which is made from polysaccharide chains cross-linked by peptides containing D-amino acids. Bacterial cell walls are different from the cell walls of plants and fungi, which are made of cellulose and chitin, respectively. The cell wall of bacteria is also distinct from that of Archaea, which do not contain peptidoglycan. The cell wall is essential to the survival of many bacteria, and the antibiotic penicillin is able to kill bacteria by inhibiting a step in the synthesis of peptidoglycan.

There are broadly speaking two different types of cell wall in bacteria, a thick one in the gram-positives and a thinner one in the gram-negatives. The names originate from the reaction of cells to the Gram stain, a long-standing test for the classification of bacterial species.

Gram-positive bacteria possess a thick cell wall containing many layers of peptidoglycan and *teichoic acids*. In contrast, *gram-negative bacteria* have a relatively thin cell wall consisting of a few layers of peptidoglycan surrounded by a second lipid membrane containing *lipopolysaccharides* and lipoproteins. Lipopolysaccharides, also called *endotoxins*, are composed of polysaccharides and *lipid A* that is responsible for much of the toxicity of gram-negative bacteria. Most bacteria have the gram-negative cell wall, and only the Firmicutes and Actinobacteria have the alternative gram-positive arrangement. These two groups were previously known as the low G+C and high G+C gram-positive bacteria, respectively. These differences in structure can produce differences in antibiotic susceptibility; for instance, vancomycin can kill only gram-positive bacteria and is ineffective against gram-negative pathogens, such as *Haemophilus influenzae* or *Pseudomonas aeruginosa*. If the bacterial cell wall is entirely removed, it is called a *protoplast*, whereas if it is partially removed, it is called a *spheroplast*. β-Lactam antibiotics, such as penicillin, inhibit the formation of peptidoglycan cross-links in the bacterial cell wall. The enzyme lysozyme, found in human tears, also digests the cell wall of bacteria and is the body's main defence against eye infections.

Acid-fast bacteria, such as *Mycobacteria*, are resistant to decolorisation by acids during staining procedures. The high mycolic acid content of *Mycobacteria*, is responsible for the staining pattern of poor absorption followed by high retention. The most common staining technique used to identify acid-fast bacteria is the Ziehl-Neelsen stain or acid-fast stain, in which the acid-fast bacilli are stained bright-red and stand out clearly against a blue background. *L-form bacteria* are strains of bacteria that lack cell walls. The main pathogenic bacteria in this class is *Mycoplasma*.

In many bacteria, an *S-layer* of rigidly arrayed protein molecules covers the outside of the cell. This layer provides chemical and physical protection for the cell surface and can act as a macro-molecular diffusion barrier. S-layers have diverse but mostly poorly understood functions, but are known to act as virulence factors in *Campylobacter* and contain surface enzymes in *Bacillus stearothermophilus*.

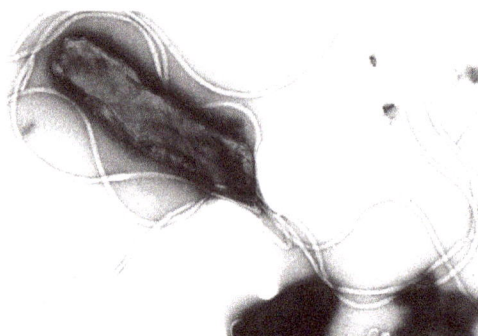

Helicobacter pylori electron micrograph, showing multiple flagella on the cell surface

Flagella are rigid protein structures, about 20 nanometres in diameter and up to 20 micrometres in length, that are used for motility. Flagella are driven by the energy released by the transfer of ions down an electrochemical gradient across the cell membrane.

Fimbriae (sometimes called "attachment pili") are fine filaments of protein, usually 2–10 nanometres in diameter and up to several micrometres in length. They are distributed over the surface of the cell, and resemble fine hairs when seen under the electron microscope. Fimbriae are believed to be involved in attachment to solid surfaces or to other cells, and are essential for the virulence of some bacterial pathogens. *Pili* (*sing.* pilus) are cellular appendages, slightly larger than fimbriae, that can transfer genetic material between bacterial cells in a process called conjugation where they are called *conjugation pili* or "sex pili". They can also generate movement where they are called *type IV pili*.

Glycocalyx are produced by many bacteria to surround their cells, and vary in structural complexity: ranging from a disorganised *slime layer* of extra-cellular polymer to a highly structured *capsule*. These structures can protect cells from engulfment by eukaryotic cells such as macrophages (part of the human immune system). They can also act as antigens and be involved in cell recognition, as well as aiding attachment to surfaces and the formation of biofilms.

The assembly of these extracellular structures is dependent on bacterial secretion systems. These transfer proteins from the cytoplasm into the periplasm or into the environment around the cell. Many types of secretion systems are known and these structures are often essential for the virulence of pathogens, so are intensively studied.

Endospores

Certain genera of gram-positive bacteria, such as *Bacillus*, *Clostridium*, *Sporohalobacter*, *Anaerobacter*, and *Heliobacterium*, can form highly resistant, dormant structures called *endospores*. In almost all cases, one endospore is formed and this is not a reproductive process, although *Anaerobacter* can make up to seven endospores in a single cell. Endospores have a central core of cytoplasm containing DNA and ribosomes surrounded by a cortex layer and protected by an imperme-able and rigid coat. Dipicolinic acid is a chemical compound that composes 5% to 15% of the dry weight of bacterial spores. It is implicated as responsible for the heat resistance of the endospore.

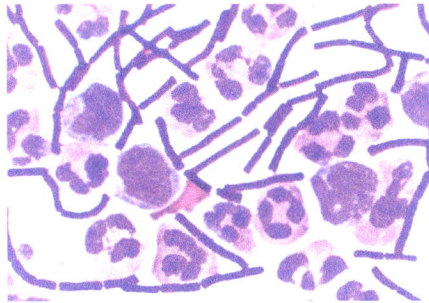

Bacillus anthracis (stained purple) growing in cerebrospinal fluid

Endospores show no detectable metabolism and can survive extreme physical and chemical stress-es, such as high levels of UV light, gamma radiation, detergents, disinfectants, heat, freezing, pres-sure, and desiccation. In this dormant state, these organisms may remain viable for millions of years, and endospores even allow bacteria to survive exposure to the vacuum and radiation in space. According to scientist Dr. Steinn Sigurdsson, "There are viable bacterial spores that have been found that are 40 million years old on Earth — and we know they're very hardened to radia-tion." Endospore-forming bacteria can also cause disease: for example, anthrax can be contracted by the inhalation of *Bacillus anthracis* endospores, and contamination of deep puncture wounds with *Clostridium tetani* endospores causes tetanus.

Metabolism

Bacteria exhibit an extremely wide variety of metabolic types. The distribution of metabolic traits within a group of bacteria has traditionally been used to define their taxonomy, but these traits often do not correspond with modern genetic classifications. Bacterial metabolism is classified into nutritional groups on the basis of three major criteria: the kind of energy used for growth, the source of carbon, and the electron donors used for growth. An additional criterion of respiratory microorganisms are the electron acceptors used for aerobic or anaerobic respiration.

Nutritional types in bacterial metabolism			
Nutritional type	Source of energy	Source of carbon	Examples
Phototrophs	Sunlight	Organic compounds (photoheterotrophs) or carbon fixation (photoautotrophs)	Cyanobacteria, Green sulfur bacteria, Chloroflexi, or Purple bacteria
Lithotrophs	Inorganic compounds	Organic compounds (lithoheterotrophs) or carbon fixation (lithoautotrophs)	Thermodesulfobacteria, Hydrogenophilaceae, or Nitrospirae
Organotrophs	Organic compounds	Organic compounds (chemoheterotrophs) or carbon fixation (chemoautotrophs)	Bacillus, Clostridium or Enterobacteriaceae

Carbon metabolism in bacteria is either *heterotrophic*, where organic carbon compounds are used as carbon sources, or *autotrophic*, meaning that cellular carbon is obtained by fixing carbon dioxide. Heterotrophic bacteria include parasitic types. Typical autotrophic bacteria are phototrophic cyanobacteria, green sulfur-bacteria and some purple bacteria, but also many chemolithotrophic species, such as nitrifying or sulfur-oxidising bacteria. Energy metabolism of bacteria is either based on *phototrophy*, the use of light through photosynthesis, or based on *chemotrophy*, the use of chemical substances for energy, which are mostly oxidised at the expense of oxygen or alternative electron acceptors (aerobic/anaerobic respiration).

Filaments of photosynthetic cyanobacteria

Bacteria are further divided into *lithotrophs* that use inorganic electron donors and *organotrophs* that use organic compounds as electron donors. Chemotrophic organisms use the respective electron donors for energy conservation (by aerobic/anaerobic respiration or fermentation) and biosynthetic reactions (e.g., carbon dioxide fixation), whereas phototrophic organisms use them only for biosynthetic purposes. Respiratory organisms use chemical compounds as a source of energy by taking electrons from the reduced substrate and transferring them to a terminal electron acceptor in a redox reaction. This reaction releases energy that can be used to synthesise ATP and drive metabolism. In *aerobic organisms*, oxygen is used as the electron acceptor. In *anaerobic organisms* other inorganic compounds, such as nitrate, sulfate or carbon dioxide are used as electron acceptors. This leads to the ecologically important processes of denitrification, sulfate reduction, and acetogenesis, respectively.

Another way of life of chemotrophs in the absence of possible electron acceptors is fermentation, wherein the electrons taken from the reduced substrates are transferred to oxidised intermediates to generate reduced fermentation products (e.g., lactate, ethanol, hydrogen, butyric acid). Fermentation is possible, because the energy content of the substrates is higher than that of the products, which allows the organisms to synthesise ATP and drive their metabolism.

These processes are also important in biological responses to pollution; for example, sulfate-reducing bacteria are largely responsible for the production of the highly toxic forms of mercury (methyl- and dimethylmercury) in the environment. Non-respiratory anaerobes use fermentation to generate energy and reducing power, secreting metabolic by-products (such as ethanol in brewing) as waste. Facultative anaerobes can switch between fermentation and different terminal electron acceptors depending on the environmental conditions in which they find themselves.

Lithotrophic bacteria can use inorganic compounds as a source of energy. Common inorganic electron donors are hydrogen, carbon monoxide, ammonia (leading to nitrification), ferrous iron and other reduced metal ions, and several reduced sulfur compounds. In unusual circumstances, the gas methane can be used by methanotrophic bacteria as both a source of electrons and a substrate for carbon anabolism. In both aerobic phototrophy and chemolithotrophy, oxygen is used as a terminal electron acceptor, whereas under anaerobic conditions inorganic compounds are used instead. Most lithotrophic organisms are autotrophic, whereas organotrophic organisms are heterotrophic.

In addition to fixing carbon dioxide in photosynthesis, some bacteria also fix nitrogen gas (nitrogen fixation) using the enzyme nitrogenase. This environmentally important trait can be found in bacteria of nearly all the metabolic types listed above, but is not universal.

Regardless of the type of metabolic process they employ, the majority of bacteria are able to take in raw materials only in the form of relatively small molecules, which enter the cell by diffusion or through molecular channels in cell membranes. The Planctomycetes are the exception (as they are in possessing membranes around their nuclear material). It has recently been shown that *Gemmata obscuriglobus* is able to take in large molecules via a process that in some ways resembles endocytosis, the process used by eukaryotic cells to engulf external items.

Growth and Reproduction

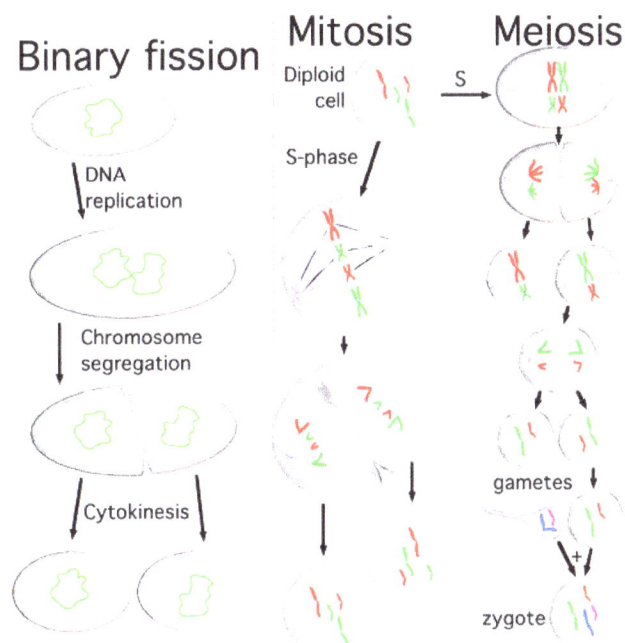

Many bacteria reproduce through binary fission, which is compared to mitosis and meiosis in this image.

Unlike in multicellular organisms, increases in cell size (cell growth) and reproduction by cell division are tightly linked in unicellular organisms. Bacteria grow to a fixed size and then reproduce through *binary fission*, a form of asexual reproduction. Under optimal conditions, bacteria can grow and divide extremely rapidly, and bacterial populations can double as quickly as every 9.8 minutes. In cell division, two identical clone daughter cells are produced. Some bacteria, while still reproducing asexually, form more complex reproductive structures that help disperse the newly formed daughter cells. Examples include fruiting body formation by *Myxobacteria* and aerial hyphae formation by *Streptomyces*, or budding. Budding involves a cell forming a protrusion that breaks away and produces a daughter cell.

A colony of *Escherichia coli*

In the laboratory, bacteria are usually grown using solid or liquid media. Solid *growth media*, such as agar plates, are used to isolate pure cultures of a bacterial strain. However, liquid growth media are used when measurement of growth or large volumes of cells are required. Growth in stirred liquid media occurs as an even cell suspension, making the cultures easy to divide and transfer, although isolating single bacteria from liquid media is difficult. The use of selective media (media with specific nutrients added or deficient, or with antibiotics added) can help identify specific organisms.

Most laboratory techniques for growing bacteria use high levels of nutrients to produce large amounts of cells cheaply and quickly. However, in natural environments, nutrients are limited, meaning that bacteria cannot continue to reproduce indefinitely. This nutrient limitation has led the evolution of different growth strategies. Some organisms can grow extremely rapidly when nutrients become available, such as the formation of algal (and cyanobacterial) blooms that often occur in lakes during the summer. Other organisms have adaptations to harsh environments, such as the production of multiple antibiotics by *Streptomyces* that inhibit the growth of competing microorganisms. In nature, many organisms live in communities (e.g., biofilms) that may allow for increased supply of nutrients and protection from environmental stresses. These relationships can be essential for growth of a particular organism or group of organisms (syntrophy).

Bacterial growth follows four phases. When a population of bacteria first enter a high-nutrient environment that allows growth, the cells need to adapt to their new environment. The first phase of growth is the *lag phase*, a period of slow growth when the cells are adapting to the high-nutrient environment and preparing for fast growth. The lag phase has high biosynthesis rates, as proteins necessary for rapid growth are produced. The second phase of growth is the *log phase*, also known as the *logarithmic or exponential phase*. The log phase is marked by rapid exponential growth. The rate at which cells grow during this phase is known as the *growth rate* (k), and the time it takes the cells to double is known as the *generation time* (g). During log phase, nutrients are metabolised at maximum speed until one of the nutrients is depleted and starts limiting growth. The third phase of growth is the *stationary phase* and is caused by depleted nutrients. The cells reduce their metabolic activity and consume non-essential cellular proteins. The stationary phase is a transition from rapid growth to a stress response state and there is increased expression of genes involved in DNA repair, antioxidant metabolism and nutrient transport. The final phase is the *death phase* where the bacteria run out of nutrients and die.

Genomes

The genomes of thousands of bacterial species have been sequenced, with at least 9,000 sequences completed and more than 42,000 left as "permanent" drafts (as of Sep 2016).

Most bacteria have a single circular chromosome that can range in size from only 160,000 base pairs in the endosymbiotic bacteria *Candidatus Carsonella ruddii*, to 12,200,000 base pairs in the soil-dwelling bacteria *Sorangium cellulosum*. The genes in bacterial genomes are usually a single continuous stretch of DNA and although several different types of introns do exist in bacteria, these are much rarer than in eukaryotes. Some bacteria, including the Spirochaetes of the genus *Borrelia* are a notable exception to this arrangement. *Borrelia burgdorferi*, the cause of Lyme disease, contains a single linear chromosome and several linear and circular plasmids.

Plasmids are small extra-chromosomal DNAs that may contain genes for antibiotic resistance or virulence factors. Plasmids replicate independently of chromosomes, so it is possible that plasmids could be lost in bacterial cell division. Against this possibility is the fact that a single bacterium can contain hundreds of copies of a single plasmid.

Genetics

Bacteria, as asexual organisms, inherit identical copies of their parent's genes (i.e., they are clonal). However, all bacteria can evolve by selection on changes to their genetic material DNA caused by genetic recombination or mutations. Mutations come from errors made during the replication of DNA or from exposure to mutagens. Mutation rates vary widely among different species of bacteria and even among different clones of a single species of bacteria. Genetic changes in bacterial genomes come from either random mutation during replication or "stress-directed mutation", where genes involved in a particular growth-limiting process have an increased mutation rate.

DNA Transfer

Some bacteria also transfer genetic material between cells. This can occur in three main ways. First, bacteria can take up exogenous DNA from their environment, in a process called *transformation*. Genes can also be transferred by the process of *transduction*, when the integration of a bacteriophage introduces foreign DNA into the chromosome. The third method of gene transfer is *conjugation*, whereby DNA is transferred through direct cell contact.

Transduction of bacterial genes by bacteriophage appears to be a consequence of infrequent errors during intracellular assembly of virus particles, rather than a bacterial adaptation. Conjugation, in the much-studied E. coli system is determined by plasmid genes, and is an adaptation for transferring copies of the plasmid from one bacterial host to another. It is seldom that a conjugative plasmid integrates into the host bacterial chromosome, and subsequently transfers part of the host bacterial DNA to another bacterium. Plasmid-mediated transfer of host bacterial DNA also appears to be an accidental process rather than a bacterial adaptation.

Transformation, unlike transduction or conjugation, depends on numerous bacterial gene products that specifically interact to perform this complex process, and thus transformation is clearly a bacterial adaptation for DNA transfer. In order for a bacterium to bind, take up and recombine donor DNA into its own chromosome, it must first enter a special physiological state termed competence. In *Bacillus subtilis*, about 40 genes are required for the development of competence. The length of DNA transferred during *B. subtilis* transformation can be between a third of a chromosome up to the whole chromosome. Transformation appears to be common among bacterial species, and thus far at least 60 species are known to have the natural ability to become competent for transformation. The development of competence in nature is usually associated with stressful environmental conditions, and seems to be an adaptation for facilitating repair of DNA damage in recipient cells.

In ordinary circumstances, transduction, conjugation, and transformation involve transfer of DNA between individual bacteria of the same species, but occasionally transfer may occur between individuals of different bacterial species and this may have significant consequences, such as the transfer of antibiotic resistance. In such cases, gene acquisition from other bacteria or the environment is called *horizontal gene transfer* and may be common under natural conditions. Gene transfer is particularly important in antibiotic resistance as it allows the rapid transfer of resistance genes between different pathogens.

Bacteriophages

Bacteriophages are viruses that infect bacteria. Many types of bacteriophage exist, some simply infect and lyse their host bacteria, while others insert into the bacterial chromosome. A bacteriophage can contain genes that contribute to its host's phenotype: for example, in the evolution of *Escherichia coli* O157:H7 and *Clostridium botulinum*, the toxin genes in an integrated phage converted a harmless ancestral bacterium into a lethal pathogen. Bacteria resist phage infection through restriction modification systems that degrade foreign DNA, and a system that uses CRISPR sequences to retain fragments of the genomes of phage that the bacteria have come into contact with in the past, which allows them to block virus replication through a form of RNA interference. This CRISPR system provides bacteria with acquired immunity to infection.

Behaviour

Secretion

Bacteria frequently secrete chemicals into their environment in order to modify it favourably. The secretions are often proteins and may act as enzymes that digest some form of food in the environment.

Bioluminescence

A few bacteria have chemical systems that generate light. This bioluminescence often occurs in bacteria that live in association with fish, and the light probably serves to attract fish or other large animals.

Multicellularity

Bacteria often function as multicellular aggregates known as biofilms, exchanging a variety of molecular signals for inter-cell communication, and engaging in coordinated multicellular behaviour.

The communal benefits of multicellular cooperation include a cellular division of labour, accessing resources that cannot effectively be used by single cells, collectively defending against antagonists, and optimising population survival by differentiating into distinct cell types. For example, bacteria in biofilms can have more than 500 times increased resistance to antibacterial agents than individual "planktonic" bacteria of the same species.

One type of inter-cellular communication by a molecular signal is called quorum sensing, which serves the purpose of determining whether there is a local population density that is sufficiently high that it is productive to invest in processes that are only successful if large numbers of similar organisms behave similarly, as in excreting digestive enzymes or emitting light.

Quorum sensing allows bacteria to coordinate gene expression, and enables them to produce, release and detect autoinducers or pheromones which accumulate with the growth in cell population.

Movement

Many bacteria can move using a variety of mechanisms: flagella are used for swimming through fluids; bacterial gliding and twitching motility move bacteria across surfaces; and changes of buoyancy allow vertical motion.

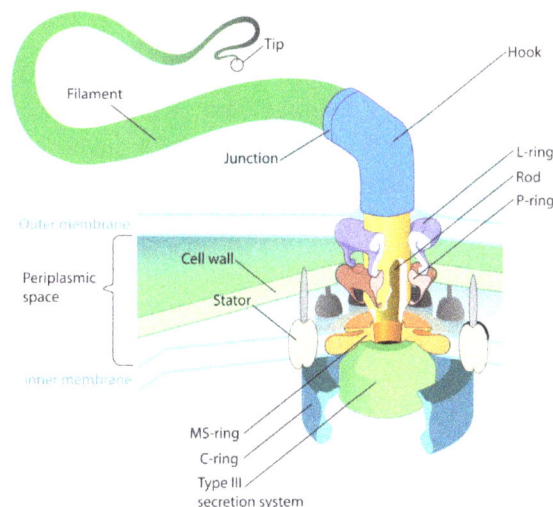

Flagellum of gram-negative bacteria. The base drives the rotation of the hook and filament.

Swimming bacteria frequently move near 10 body lengths per second and a few as fast as 100. This makes them at least as fast as fish, on a relative scale.

In bacterial gliding and twitching motility, bacteria use their *type IV pili* as a grappling hook, repeatedly extending it, anchoring it and then retracting it with remarkable force (>80 pN).

"Our observations redefine twitching motility as a rapid, highly organized mechanism of bacterial translocation by which Pseudomonas aeruginosa can disperse itself over large areas to colonize new territories. It is also now clear, both morphologically and genetically, that twitching motility and social gliding motility, such as occurs in Myxococcus xanthus, are essentially the same process."

— "A re-examination of twitching motility in Pseudomonas aeruginosa" – Semmler, Whitchurch & Mattick (1999)

Flagella are semi-rigid cylindrical structures that are rotated and function much like the propeller on a ship. Objects as small as bacteria operate a low Reynolds number and cylindrical forms are more efficient than the flat, paddle-like, forms appropriate at human-size scale.

Bacterial species differ in the number and arrangement of flagella on their surface; some have a single flagellum (*monotrichous*), a flagellum at each end (*amphitrichous*), clusters of flagella at the poles of the cell (*lophotrichous*), while others have flagella distributed over the entire surface of the cell (*peritrichous*). The bacterial flagella is the best-understood motility structure in any organism and is made of about 20 proteins, with approximately another 30 proteins required for its regulation and assembly. The flagellum is a rotating structure driven by a reversible motor at the base that uses the electrochemical gradient across the membrane for power. This motor drives the motion of the filament, which acts as a propeller.

Many bacteria (such as *E. coli*) have two distinct modes of movement: forward movement (swimming) and tumbling. The tumbling allows them to reorient and makes their movement a three-dimensional random walk. The flagella of a unique group of bacteria, the spirochaetes, are found between two membranes in the periplasmic space. They have a distinctive helical body that twists about as it moves.

Motile bacteria are attracted or repelled by certain stimuli in behaviours called taxes: these include chemotaxis, phototaxis, energy taxis, and magnetotaxis. In one peculiar group, the myxobacteria, individual bacteria move together to form waves of cells that then differentiate to form fruiting bodies containing spores. The myxobacteria move only when on solid surfaces, unlike *E. coli*, which is motile in liquid or solid media.

Several *Listeria* and *Shigella* species move inside host cells by usurping the cytoskeleton, which is normally used to move organelles inside the cell. By promoting actin polymerisation at one pole of their cells, they can form a kind of tail that pushes them through the host cell's cytoplasm.

Classification and Identification

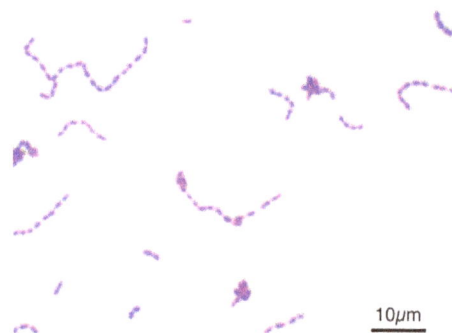

Streptococcus mutans visualised with a Gram stain

Classification seeks to describe the diversity of bacterial species by naming and grouping organisms based on similarities. Bacteria can be classified on the basis of cell structure, cellular metabolism or on differences in cell components, such as DNA, fatty acids, pigments, antigens and quinones. While these schemes allowed the identification and classification of bacterial strains, it was unclear whether these differences represented variation between distinct species or between strains of the same species. This uncertainty was due to the lack of distinctive structures in most bacteria, as well as lateral gene transfer between unrelated species. Due to lateral gene transfer, some closely related bacteria can have very different morphologies and metabolisms. To overcome this uncertainty, modern bacterial classification emphasises molecular systematics, using genetic techniques such as guanine cytosine ratio determination, genome-genome hybridisation, as well as sequencing genes that have not undergone extensive lateral gene transfer, such as the rRNA gene. Classification of bacteria is determined by publication in the International Journal of Systematic Bacteriology, and Bergey's Manual of Systematic Bacteriology. The International Committee on Systematic Bacteriology (ICSB) maintains international rules for the naming of bacteria and taxonomic categories and for the ranking of them in the International Code of Nomenclature of Bacteria.

The term "bacteria" was traditionally applied to all microscopic, single-cell prokaryotes. However, molecular systematics showed prokaryotic life to consist of two separate domains, originally called *Eubacteria* and *Archaebacteria*, but now called *Bacteria* and *Archaea* that evolved independently from an ancient common ancestor. The archaea and eukaryotes are more closely related to each other than either is to the bacteria. These two domains, along with Eukarya, are the basis of the three-domain system, which is currently the most widely used classification system in microbiolology. However, due to the relatively recent introduction of molecular systematics and a rapid increase in the number of genome sequences that are available, bacterial classification remains a changing and expanding field. For example, a few biologists argue that the Archaea and Eukaryotes evolved from gram-positive bacteria.

The identification of bacteria in the laboratory is particularly relevant in medicine, where the correct treatment is determined by the bacterial species causing an infection. Consequently, the need to identify human pathogens was a major impetus for the development of techniques to identify bacteria.

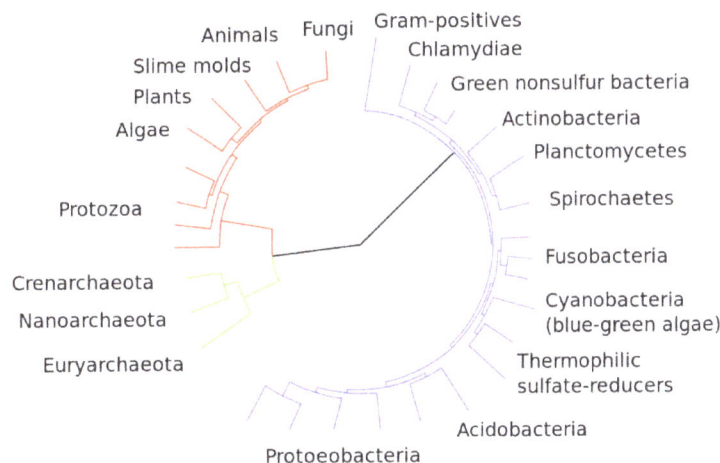

Phylogenetic tree showing the diversity of bacteria, compared to other organisms. Eukaryotes are coloured red, archaea green and bacteria blue.

The *Gram stain*, developed in 1884 by Hans Christian Gram, characterises bacteria based on the structural characteristics of their cell walls. The thick layers of peptidoglycan in the "gram-positive" cell wall stain purple, while the thin "gram-negative" cell wall appears pink. By combining morphology and Gram-staining, most bacteria can be classified as belonging to one of four groups (gram-positive cocci, gram-positive bacilli, gram-negative cocci and gram-negative bacilli). Some organisms are best identified by stains other than the Gram stain, particularly mycobacteria or *Nocardia*, which show acid-fastness on Ziehl–Neelsen or similar stains. Other organisms may need to be identified by their growth in special media, or by other techniques, such as serology.

Culture techniques are designed to promote the growth and identify particular bacteria, while restricting the growth of the other bacteria in the sample. Often these techniques are designed for specific specimens; for example, a sputum sample will be treated to identify organisms that cause pneumonia, while stool specimens are cultured on selective media to identify organisms that cause diarrhoea, while preventing growth of non-pathogenic bacteria. Specimens that are normally sterile, such as blood, urine or spinal fluid, are cultured under conditions designed to grow all possible organisms. Once a pathogenic organism has been isolated, it can be further characterised by its morphology, growth patterns (such as aerobic or anaerobic growth), patterns of hemolysis, and staining.

As with bacterial classification, identification of bacteria is increasingly using molecular methods. Diagnostics using DNA-based tools, such as polymerase chain reaction, are increasingly popular due to their specificity and speed, compared to culture-based methods. These methods also allow the detection and identification of "viable but nonculturable" cells that are metabolically active but non-dividing. However, even using these improved methods, the total number of bacterial species is not known and cannot even be estimated with any certainty. Following present classification, there are a little less than 9,300 known species of prokaryotes, which includes bacteria and archaea; but attempts to estimate the true number of bacterial diversity have ranged from 10^7 to 10^9 total species – and even these diverse estimates may be off by many orders of magnitude.

Interactions with other Organisms

Despite their apparent simplicity, bacteria can form complex associations with other organisms. These symbiotic associations can be divided into parasitism, mutualism and commensalism. Due to their small size, commensal bacteria are ubiquitous and grow on animals and plants exactly as they will grow on any other surface. However, their growth can be increased by warmth and sweat, and large populations of these organisms in humans are the cause of body odour.

Predators

Some species of bacteria kill and then consume other microorganisms, these species are called *predatory bacteria*. These include organisms such as *Myxococcus xanthus*, which forms swarms of cells that kill and digest any bacteria they encounter. Other bacterial predators either attach to their prey in order to digest them and absorb nutrients, such as *Vampirovibrio chlorellavorus*, or invade another cell and multiply inside the cytosol, such as *Daptobacter*. These predatory bacteria are thought to have evolved from saprophages that consumed dead microorganisms, through adaptations that allowed them to entrap and kill other organisms.

Mutualists

Certain bacteria form close spatial associations that are essential for their survival. One such mutualistic association, called interspecies hydrogen transfer, occurs between clusters of anaerobic bacteria that consume organic acids, such as butyric acid or propionic acid, and produce hydrogen, and methanogenic Archaea that consume hydrogen. The bacteria in this association are unable to consume the organic acids as this reaction produces hydrogen that accumulates in their surroundings. Only the intimate association with the hydrogen-consuming Archaea keeps the hydrogen concentration low enough to allow the bacteria to grow.

In soil, microorganisms that reside in the rhizosphere (a zone that includes the root surface and the soil that adheres to the root after gentle shaking) carry out nitrogen fixation, converting nitrogen gas to nitrogenous compounds. This serves to provide an easily absorbable form of nitrogen for many plants, which cannot fix nitrogen themselves. Many other bacteria are found as symbionts in humans and other organisms. For example, the presence of over 1,000 bacterial species in the normal human gut flora of the intestines can contribute to gut immunity, synthesise vitamins, such as folic acid, vitamin K and biotin, convert sugars to lactic acid, as well as fermenting complex undigestible carbohydrates. The presence of this gut flora also inhibits the growth of potentially pathogenic bacteria (usually through competitive exclusion) and these beneficial bacteria are consequently sold as probiotic dietary supplements.

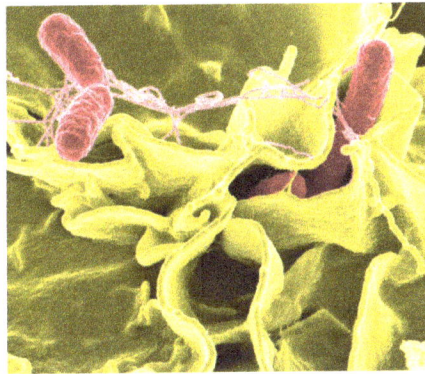

Colour-enhanced scanning electron micrograph showing *Salmonella typhimurium* (red) invading cultured human cells

Pathogens

If bacteria form a parasitic association with other organisms, they are classed as pathogens. Pathogenic bacteria are a major cause of human death and disease and cause infections such as tetanus, typhoid fever, diphtheria, syphilis, cholera, foodborne illness, leprosy and tuberculosis. A pathogenic cause for a known medical disease may only be discovered many years after, as was the case with *Helicobacter pylori* and peptic ulcer disease. Bacterial diseases are also important in agriculture, with bacteria causing leaf spot, fire blight and wilts in plants, as well as Johne's disease, mastitis, salmonella and anthrax in farm animals.

Each species of pathogen has a characteristic spectrum of interactions with its human hosts. Some organisms, such as *Staphylococcus* or *Streptococcus*, can cause skin infections, pneumonia, meningitis and even overwhelming sepsis, a systemic inflammatory response producing shock, massive vasodilation and death. Yet these organisms are also part of the normal human flora and usually exist on the skin or in the nose without causing any disease at all. Other organisms invariably cause disease in humans, such as the Rickettsia, which are obligate intracellular parasites able to grow and reproduce only within the cells of other organisms. One species of Rickettsia causes typhus, while another causes Rocky Mountain spotted fever. *Chlamydia*, another phylum of obligate intracellular parasites, contains species that can cause pneumonia, or urinary tract infection and may be involved in coronary heart disease. Finally, some species, such as *Pseudomonas aeruginosa*, *Burkholderia cenocepacia*, and *Mycobacterium avium*, are opportunistic pathogens and cause disease mainly in people suffering from immunosuppression or cystic fibrosis.

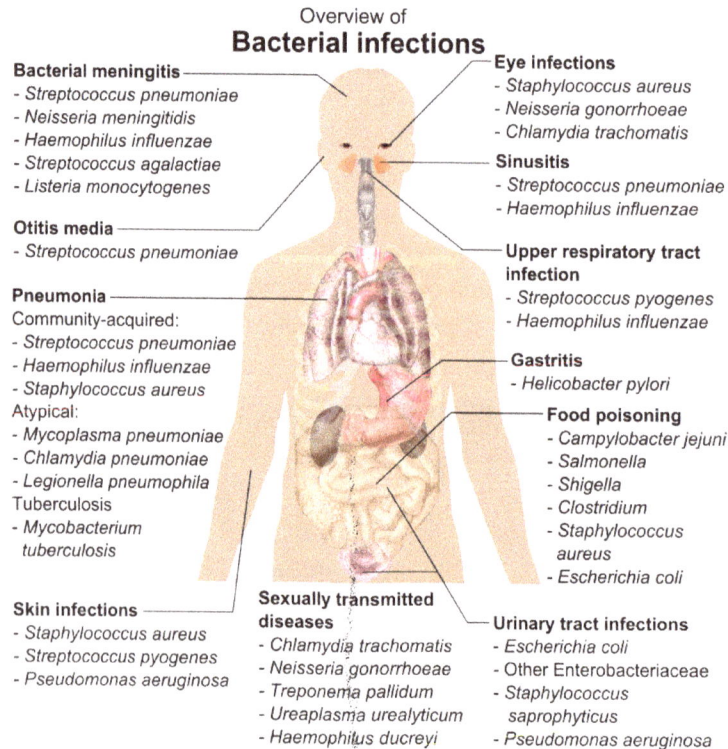

Overview of
Bacterial infections

Bacterial meningitis
- *Streptococcus pneumoniae*
- *Neisseria meningitidis*
- *Haemophilus influenzae*
- *Streptococcus agalactiae*
- *Listeria monocytogenes*

Otitis media
- *Streptococcus pneumoniae*

Pneumonia
Community-acquired:
- *Streptococcus pneumoniae*
- *Haemophilus influenzae*
- *Staphylococcus aureus*
Atypical:
- *Mycoplasma pneumoniae*
- *Chlamydia pneumoniae*
- *Legionella pneumophila*
Tuberculosis
- *Mycobacterium tuberculosis*

Skin infections
- *Staphylococcus aureus*
- *Streptococcus pyogenes*
- *Pseudomonas aeruginosa*

Sexually transmitted diseases
- *Chlamydia trachomatis*
- *Neisseria gonorrhoeae*
- *Treponema pallidum*
- *Ureaplasma urealyticum*
- *Haemophilus ducreyi*

Eye infections
- *Staphylococcus aureus*
- *Neisseria gonorrhoeae*
- *Chlamydia trachomatis*

Sinusitis
- *Streptococcus pneumoniae*
- *Haemophilus influenzae*

Upper respiratory tract infection
- *Streptococcus pyogenes*
- *Haemophilus influenzae*

Gastritis
- *Helicobacter pylori*

Food poisoning
- *Campylobacter jejuni*
- *Salmonella*
- *Shigella*
- *Clostridium*
- *Staphylococcus aureus*
- *Escherichia coli*

Urinary tract infections
- *Escherichia coli*
- Other Enterobacteriaceae
- *Staphylococcus saprophyticus*
- *Pseudomonas aeruginosa*

Overview of bacterial infections and main species involved.

Bacterial infections may be treated with antibiotics, which are classified as bacteriocidal if they kill bacteria, or bacteriostatic if they just prevent bacterial growth. There are many types of antibiotics and each class inhibits a process that is different in the pathogen from that found in the host. An example of how antibiotics produce selective toxicity are chloramphenicol and puromycin, which inhibit the bacterial ribosome, but not the structurally different eukaryotic ribosome. Antibiotics are used both in treating human disease and in intensive farming to promote animal growth, where they may be contributing to the rapid development of antibiotic resistance in bacterial populations. Infections can be prevented by antiseptic measures such as sterilising the skin prior to piercing it with the needle of a syringe, and by proper care of indwelling catheters. Surgical and dental instruments are also sterilised to prevent contamination by bacteria. Disinfectants such as bleach are used to kill bacteria or other pathogens on surfaces to prevent contamination and further reduce the risk of infection.

Significance in technology and Industry

Bacteria, often lactic acid bacteria, such as *Lactobacillus* and *Lactococcus*, in combination with yeasts and moulds, have been used for thousands of years in the preparation of fermented foods, such as cheese, pickles, soy sauce, sauerkraut, vinegar, wine and yogurt.

The ability of bacteria to degrade a variety of organic compounds is remarkable and has been used in waste processing and bioremediation. Bacteria capable of digesting the hydrocarbons in petroleum are often used to clean up oil spills. Fertiliser was added to some of the beaches in Prince William Sound in an attempt to promote the growth of these naturally occurring bacteria after the 1989 *Exxon Valdez* oil spill. These efforts were effective on beaches that were not too thickly covered in oil. Bacteria are also used for the bioremediation of industrial toxic wastes. In the chemical industry, bacteria are most important in the production of enantiomerically pure chemicals for use as pharmaceuticals or agrichemicals.

Bacteria can also be used in the place of pesticides in the biological pest control. This commonly involves *Bacillus thuringiensis* (also called BT), a gram-positive, soil dwelling bacterium. Subspecies of this bacteria are used as a Lepidopteran-specific insecticides under trade names such as Dipel and Thuricide. Because of their specificity, these pesticides are regarded as environmentally friendly, with little or no effect on humans, wildlife, pollinators and most other beneficial insects.

Because of their ability to quickly grow and the relative ease with which they can be manipulated, bacteria are the workhorses for the fields of molecular biology, genetics and biochemistry. By making mutations in bacterial DNA and examining the resulting phenotypes, scientists can determine the function of genes, enzymes and metabolic pathways in bacteria, then apply this knowledge to more complex organisms. This aim of understanding the biochemistry of a cell reaches its most complex expression in the synthesis of huge amounts of enzyme kinetic and gene expression data into mathematical models of entire organisms. This is achievable in some well-studied bacteria, with models of *Escherichia coli* metabolism now being produced and tested. This understanding of bacterial metabolism and genetics allows the use of biotechnology to bioengineer bacteria for the production of therapeutic proteins, such as insulin, growth factors, or antibodies.

Because of their importance for research in general, samples of bacterial strains are isolated and preserved in Biological Resource Centers. This ensures the availability of the strain to scientists worldwide.

History of Bacteriology

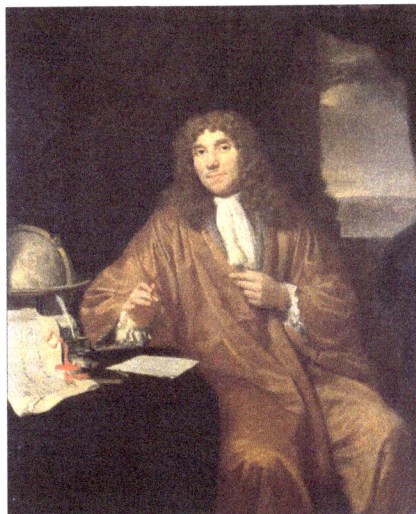

Antonie van Leeuwenhoek, the first microbiologist and the first person to observe bacteria using a microscope.

Bacteria were first observed by the Dutch microscopist Antonie van Leeuwenhoek in 1676, using a single-lens microscope of his own design. He then published his observations in a series of letters to the Royal Society of London. Bacteria were Leeuwenhoek's most remarkable microscopic discovery. They were just at the limit of what his simple lenses could make out and, in one of the most striking hiatuses in the history of science, no one else would see them again for over a century. Only then were his by-then-largely-forgotten observations of bacteria — as opposed to his famous "animalcules" (spermatozoa) — taken seriously.

Christian Gottfried Ehrenberg introduced the word "bacterium" in 1828. In fact, his *Bacterium* was a genus that contained non-spore-forming rod-shaped bacteria, as opposed to *Bacillus*, a genus of spore-forming rod-shaped bacteria defined by Ehrenberg in 1835.

Louis Pasteur demonstrated in 1859 that the growth of microorganisms causes the fermentation process, and that this growth is not due to spontaneous generation. (Yeasts and moulds, commonly associated with fermentation, are not bacteria, but rather fungi.) Along with his contemporary Robert Koch, Pasteur was an early advocate of the germ theory of disease.

Robert Koch, a pioneer in medical microbiology, worked on cholera, anthrax and tuberculosis. In his research into tuberculosis Koch finally proved the germ theory, for which he received a Nobel Prize in 1905. In *Koch's postulates*, he set out criteria to test if an organism is the cause of a disease, and these postulates are still used today.

Though it was known in the nineteenth century that bacteria are the cause of many diseases, no effective antibacterial treatments were available. In 1910, Paul Ehrlich developed the first antibiotic, by changing dyes that selectively stained *Treponema pallidum* — the spirochaete that causes syphilis — into compounds that selectively killed the pathogen. Ehrlich had been awarded a 1908 Nobel Prize for his work on immunology, and pioneered the use of stains to detect and identify bacteria, with his work being the basis of the Gram stain and the Ziehl–Neelsen stain.

A major step forward in the study of bacteria came in 1977 when Carl Woese recognised that archaea have a separate line of evolutionary descent from bacteria. This new phylogenetic taxonomy depended on the sequencing of 16S ribosomal RNA, and divided prokaryotes into two evolutionary domains, as part of the three-domain system.

Methicillin-resistant Staphylococcus aureus

Methicillin-resistant *Staphylococcus aureus* (MRSA) is a bacterium responsible for several difficult-to-treat infections in humans. MRSA is any strain of *Staphylococcus aureus* that has developed, through horizontal gene transfer and natural selection, multi- resistance to beta-lactam antibiotics, which include the penicillins (methicillin, dicloxacillin, nafcillin, oxacillin, etc.) and the cephalosporins. Strains unable to resist these antibiotics are classified as methicillin-susceptible *Staphylococcus aureus*, or MSSA. The evolution of such resistance does not cause the organism to be more intrinsically virulent than strains of *S. aureus* that have no antibiotic resistance, but resistance does make MRSA infection more difficult to treat with standard types of antibiotics and thus more dangerous.

MRSA is especially troublesome in hospitals, prisons, and nursing homes, where patients with open wounds, invasive devices, and weakened immune systems are at greater risk of nosocomial infection (hospital-acquired infection) than the general public. MRSA began as a hospital-acquired infection, but has developed limited endemic status and is now sometimes community-acquired as well as livestock-acquired. The terms HA-MRSA (healthcare-associated MRSA), CA-MRSA (community-associated MRSA) and LA-MRSA (livestock-associated) reflect this distinction.

Signs and Symptoms

S. aureus most commonly colonizes under the anterior nares (the nostrils). The rest of the respiratory tract, open wounds, intravenous catheters, and the urinary tract are also potential sites for infection. Healthy individuals may carry MRSA asymptomatically for periods ranging from a few weeks to many years. Patients with compromised immune systems are at a significantly greater risk of symptomatic secondary infection.

In most patients, MRSA can be detected by swabbing the nostrils and isolating the bacteria found inside the nostrils. Combined with extra sanitary measures for those in contact with infected patients, swab screening patients admitted to hospitals has been found to be effective in minimizing the spread of MRSA in hospitals in the United States, Denmark, Finland, and the Netherlands.

MRSA may progress substantially within 24–48 hours of initial topical symptoms. After 72 hours, MRSA can take hold in human tissues and eventually become resistant to treatment. The initial presentation of MRSA is small red bumps that resemble pimples, spider bites, or boils; they may be accompanied by fever and, occasionally, rashes. Within a few days, the bumps become larger and more painful; they eventually open into deep, pus-filled boils. About 75 percent of community-associated (CA-) MRSA infections are localized to skin and soft tissue and usually can be treated effectively. Some CA-MRSA strains display enhanced virulence, spreading more rapidly and causing illness much more severe than traditional HA-MRSA infections, and they can affect vital organs and lead to widespread infection (sepsis), toxic shock syndrome, and necrotizing pneumonia. This is thought to be due to toxins carried by CA-MRSA strains, such as PVL and PSM, though PVL was recently found not to be a factor in a study by the National Institute of Allergy and Infectious Diseases at the National Institutes of Health. It is not known why some healthy people develop CA-MRSA skin infections that are treatable while others infected with the same strain develop severe infections or die.

People are occasionally colonized with CA-MRSA and are completely asymptomatic. The most common manifestations of CA-MRSA are simple skin infections, such as impetigo, boils, abscesses, folliculitis, and cellulitis. Rarer, but more serious, manifestations can occur, such as necrotizing fasciitis and pyomyositis (most commonly found in the tropics), necrotizing pneumonia, and infective endocarditis (which affects the valves of the heart), and bone and joint infections. CA-MRSA often results in abscess formation that requires incision and drainage. Before the spread of MRSA into the community, abscesses were not considered contagious, because infection was assumed to require violation of skin integrity and the introduction of staphylococci from normal skin colonization. However, newly emerging CA-MRSA is transmissible (similar, but with very important differences) from HA-MRSA. CA-MRSA is less likely than other forms of MRSA to cause cellulitis.

Risk Factors

Some of the populations at risk:

- People who are frequently in crowded places, especially with shared equipment and skin-to-skin contact

- People with weak immune systems (HIV/AIDS, lupus, or cancer sufferers; transplant recipients, severe asthmatics, etc.)

- Diabetics

- Intravenous drug users

- Users of quinolone antibiotics

- The elderly

- School children sharing sports and other equipment

- College students living in dormitories

- Women with frequent urinary tract or kidney infections due to infections in the bladder

- People staying or working in a health care facility for an extended period of time

- People who spend time in coastal waters where MRSA is present, such as some beaches in Florida and the west coast of the United States

- People who spend time in confined spaces with other people, including occupants of homeless shelters and warming centers, prison inmates, military recruits in basic training, and individuals who spend considerable time in changing rooms or gyms

- Veterinarians, livestock handlers, and pet owners

Hospital Patients

Many MRSA infections occur in hospitals and healthcare facilities. Infections occurring in this manner are known as healthcare acquired MRSA (HA-MRSA). The rates of MRSA infection are also increased in hospitalized patients who are treated with quinolones. Healthcare provider-to-patient transfer is common, especially when healthcare providers move from patient to patient without performing necessary hand-washing techniques between patients. Online tools predicting probability of nasal carriage in hospital admissions are available.

Prison Inmates, Military Recruits, and the Homeless

Prisons, military barracks, and homeless shelters can be crowded and confined, and poor hygiene practices may proliferate, thus putting inhabitants at increased risk of contracting MRSA. Cases of MRSA in such populations were first reported in the United States, and then in Canada. The earliest reports were made by the Center for Disease Control (CDC) in US state prisons. Subsequent reports of a massive rise in skin and soft tissue infections were reported by the CDC in the Los Angeles County Jail system in 2001, and this has continued. Pan et al. reported on the changing epidemiology of MRSA skin infection in the San Francisco County Jail, noting MRSA accounted for more than 70% of *S. aureus* infection in the jail by 2002. Lowy and colleagues reported on frequent MRSA skin infections in New York state prisons. Two reports on inmates in Maryland have demonstrated frequent colonization with MRSA.

In the news media, hundreds of reports of MRSA outbreaks in prisons appeared between 2000 and 2008. For example, in February 2008, the Tulsa County jail in Oklahoma started treating an average of 12 *S. aureus* cases per month. A report on skin and soft tissue infections in the Cook County jail in Chicago in 2004–05 demonstrated MRSA was the most common cause of these infections among cultured lesions, and few risk factors were more strongly associated with MRSA infections than infections caused by methicillin-susceptible *S. aureus*. In response to these and many other reports on MRSA infections among incarcerated and recently incarcerated persons, the Federal Bureau of Prisons has released guidelines for the management and control of the infections, although few studies provide an evidence base for these guidelines.

Livestock

Cases of MRSA have increased in livestock animals. CC398, a new variant of MRSA, has emerged in animals and is found in intensively reared production animals (primarily pigs, but also cattle and poultry), where it can be transmitted to humans as LA-MRSA (livestock-associated MRSA). Though dangerous to humans, CC398 is often asymptomatic in food-producing animals. In a single study conducted in Denmark, MRSA was shown to originate in livestock and spread to humans, though the MRSA strain may have originated in humans and was transmitted to livestock.

A 2011 study reported 47% of the meat and poultry sold in surveyed U.S. grocery stores was contaminated with S. aureus, and of those, 52% — or 24.4% of the total — were resistant to at least three classes of antibiotics. "Now we need to determine what this means in terms of risk to the consumer," said Dr. Keim, a co-author of the paper. Some samples of commercially sold meat products in Japan were also found to harbor MRSA strains.

An investigation of 100 pig-meat samples purchased from major UK retailers conducted by the Guardian in 2015 showed that some 10% of the samples were contaminated.

Athletes

Locker rooms, gyms, and related athletic facilities offer potential sites for MRSA contamination and infection. A study linked MRSA to the abrasions caused by artificial turf. Three studies by the Texas State Department of Health found the infection rate among football players was 16 times the national average. In October 2006, a high-school football player was temporarily paralyzed from MRSA-infected turf burns. His infection returned in January 2007 and required three surgeries to remove infected tissue, as well as three weeks of hospital stay. In 2013, Lawrence Tynes, Carl Nicks, and Johnthan Banks of the Tampa Bay Buccaneers were diagnosed with MRSA. Tynes and Nicks apparently did not contract the infection from each other, but it is unknown if Banks contracted it from either individual. In 2015, Los Angeles Dodgers' infielder Justin Turner was infected while the team visited the New York Mets. In October 2015, New York Giants tight end Daniel Fells was hospitalized with a serious MRSA infection.

Children

MRSA is becoming a critical problem in pediatric settings; recent studies found 4.6% of patients in U.S. health-care facilities, (presumably) including hospital nurseries, were infected or colonized with MRSA. Children (and adults, as well) who come in contact with day-care centers, playgrounds, locker rooms, camps, dormitories, classrooms and other school settings, and gyms and workout facilities are at higher risk of getting MRSA. Parents should be especially cautious of children who participate in activities where sports equipment is shared, such as football helmets and uniforms.

Diagnosis

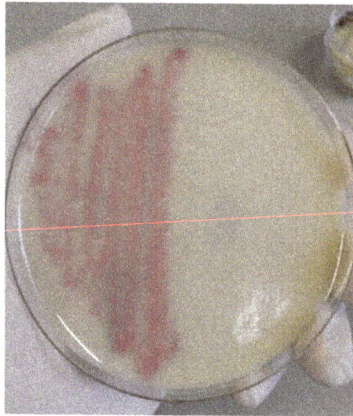

A selective and differential chromogenic medium for the qualitative direct detection of MRSA.

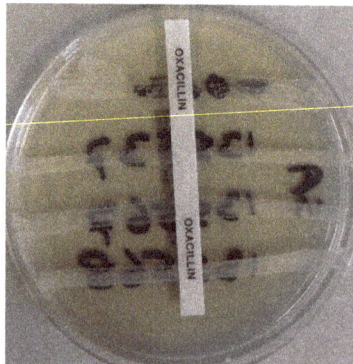

The MRSA resistance to oxacillin being tested, the top s. aureus isolate is control and sensitive to oxacillin, the other three isolates are MRSA positive

Diagnostic microbiology laboratories and reference laboratories are key for identifying outbreaks of MRSA. Faster techniques for identifying and characterizing MRSA have recently been developed. Normally, the bacterium must be cultured from blood, urine, sputum, or other body-fluid samples, and in sufficient quantities to perform confirmatory tests early-on. Still, because no quick and easy method exists to diagnose MRSA, initial treatment of the infection is often based upon 'strong suspicion' and techniques by the treating physician; these include quantitative PCR procedures, which are employed in clinical laboratories for quickly detecting and identifying MRSA strains.

Another common laboratory test is a rapid latex agglutination test that detects the PBP2a protein.

PBP2a is a variant penicillin-binding protein that imparts the ability of *S. aureus* to be resistant to oxacillin.

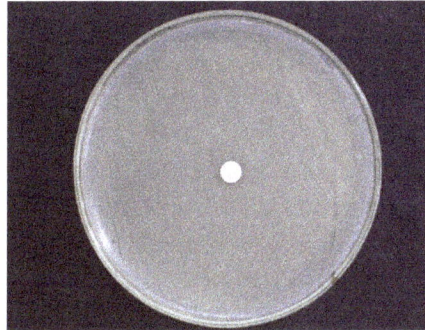

Mueller Hinton agar showing MRSA resistant to oxacillin disk

Genetics

Antimicrobial resistance is genetically based; resistance is mediated by the acquisition of extra-chromosomal genetic elements containing resistance genes. Examples include plasmids, transposable genetic elements, and genomic islands, which are transferred between bacteria through horizontal gene transfer. A defining characteristic of MRSA is its ability to thrive in the presence of penicillin-like antibiotics, which normally prevent bacterial growth by inhibiting synthesis of cell wall material. This is due to a resistance gene, *mecA*, which stops β-lactam antibiotics from inactivating the enzymes (transpeptidases) critical for cell wall synthesis.

SCCmec

Staphylococcal cassette chromosome *mec* (SCC*mec*) is a genomic island of unknown origin containing the antibiotic resistance gene *mecA*. SCC*mec* contains additional genes beyond *mecA*, including the cytolysin gene *psm-mec*, which may suppress virulence in HA-acquired MRSA strains. SCC*mec* also contains *ccrA* and *ccrB*; both genes encode recombinases that mediate the site-specific integration and excision of the SCC*mec* element from the *S. aureus* chromosome. Currently, six unique SCC*mec* types ranging in size from 21–67 kb have been identified; they are designated types I-VI and are distinguished by variation in *mec* and *ccr* gene complexes. Owing to the size of the SCC*mec* element and the constraints of horizontal gene transfer, a limited number of clones is thought to be responsible for the spread of MRSA infections.

Different SCC*mec* genotypes confer different microbiological characteristics, such as different antimicrobial resistance rates. Different genotypes are also associated with different types of infections. Types I-III SCC*mec* are large elements that typically contain additional resistance genes and are characteristically isolated from HA-MRSA strains. Conversely, CA-MRSA is associated with types IV and V, which are smaller and lack resistance genes other than *mecA*.

mecA

mecA is a biomarker gene responsible for resistance to methicillin and other β-lactam antibiotics. After acquisition of *mecA*, the gene must be integrated and localized in the S. aureus chromosome. *mecA* encodes penicillin-binding protein 2a (PBP2a), which differs from other penicillin-binding proteins as its active site does not bind methicillin or other β-lactam antibiotics. As such, PBP2a can continue to catalyze the transpeptidation reaction required for peptidoglycan cross-linking, enabling cell wall synthesis in the presence of antibiotics. As a consequence of the inability of PBP2a to interact with β-lactam moieties, acquisition of *mecA* confers resistance to all β-lactam antibiotics in addition to methicillin.

mecA is under the control of two regulatory genes, *mecI* and *mecR1*. MecI is usually bound to the *mecA* promoter and functions as a repressor. In the presence of a β-lactam antibiotic, MecR1 initiates a signal transduction cascade that leads to transcriptional activation of *mecA*. This is achieved by MecR1-mediated cleavage of MecI, which alleviates MecI repression. *mecA* is further controlled by two co-repressors, BlaI and BlaR1. *blaI* and *blaR1* are homologous to *mecI* and *mecR1*, respectively, and normally function as regulators of *blaZ*, which is responsible for penicillin resistance. The DNA sequences bound by MecI and BlaI are identical; therefore, BlaI can also bind the *mecA* operator to repress transcription of *mecA*.

Arginine Catabolic Mobile Element

The arginine catabolic mobile element (ACME) is a virulence factor present in many MRSA strains but not prevalent in MSSA. SpeG-positive ACME compensates for the polyamine hypersensitivity of *S. aureus* and facilitates stable skin colonization, wound infection, and person-to-person transmission.

Strains

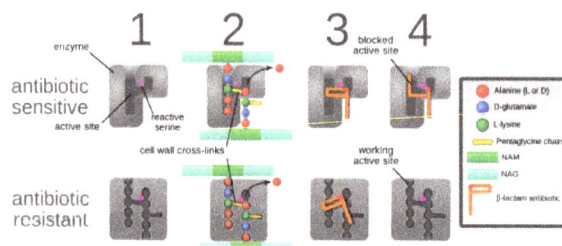

Diagram depicting antibiotic resistance through alteration of the antibiotic's target site, modeled after MRSA's resistance to penicillin. Beta-lactam antibiotics permanently inactivate PBP enzymes, which are essential for bacterial life, by permanently binding to their active sites. Some forms of MRSA, however, express a PBP that will not allow the antibiotic into their active site.

Acquisition of SCC*mec* in methicillin-sensitive staphylococcus aureus (*MSSA*) gives rise to a number of genetically different MRSA lineages. These genetic variations within different MRSA strains possibly explain the variability in virulence and associated MRSA infections. The first MRSA strain, ST250 MRSA-1 originated from SCC*mec* and ST250-MSSA integration. Historically, major MRSA clones: ST2470-MRSA-I, ST239-MRSA-III, ST5-MRSA-II, and ST5-MRSA-IV were responsible for causing hospital-acquired MRSA (HA-MRSA) infections. ST239-MRSA-III, known as the Brazilian clone, was highly transmissible compared to others and distributed in Argentina, Czech Republic, and Portugal.

In the UK, the most common strains of MRSA are EMRSA15 and EMRSA16. EMRSA16 is the best described epidemiologically: it originated in Kettering, England, and the full genomic sequence of this strain has been published. EMRSA16 has been found to be identical to the ST36:USA200 strain, which circulates in the United States, and to carry the SCC*mec* type II, enterotoxin A and toxic shock syndrome toxin 1 genes. Under the new international typing system, this strain is now called MRSA252. EMRSA 15 is also found to be one of the common MRSA strains in Asia. Other common strains include ST5:USA100 and EMRSA 1. These strains are genetic characteristics of HA-MRSA.

It is not entirely certain why some strains are highly transmissible and persistent in healthcare facilities. One explanation is the characteristic pattern of antibiotic susceptibility. Both the EMRSA15 and EMRSA16 strains are resistant to erythromycin and ciprofloxacin. It is known that *Staphylococcus aureus* can survive intracellularly, for example in the nasal mucosa and in the tonsil tissue. Erythromycin and ciprofloxacin are precisely the antibiotics that best penetrate intracellularly; it may be that these strains of *S. aureus* are therefore able to exploit an intracellular niche.

Community-acquired MRSA (CA-MRSA) strains emerged in late 1990 to 2000, infecting healthy people who had not been in contact with health care facilities. A later study that analyzed data from more than 300 microbiology labs associated with hospitals all over the United States have found a seven-fold increase, jumping from 3.6% of all MRSA infections to 28.2%, in the proportion of community-associated strains of MRSA between 1999 and 2006. Researchers suggest that CA-MRSA did not evolve from the HA-MRSA. This is further proven by molecular typing of CA-MRSA strains and genome comparison between CA-MRSA and HA-MRSA, which indicate that novel MRSA strains integrated SCC*mec* into MSSA separately on its own. By mid 2000, CA-MRSA was introduced into the health care systems and distinguishing CA-MRSA from HA-MRSA became a difficult process. Community-acquired MRSA (CA-MRSA) is more easily treated and more virulent than hospital-acquired MRSA (HA-MRSA). The genetic mechanism for the enhanced virulence in CA-MRSA remains an active area of research. Especially the Panton–Valentine leukocidin (PVL) genes are of interest because they are a unique feature of CA-MRSA.

In the United States, most cases of CA-MRSA are caused by a CC8 strain designated ST8:USA300, which carries SCC*mec* type IV, Panton–Valentine leukocidin, PSM-alpha and enterotoxins Q and K, and ST1:USA400. The ST8:USA300 strain results in skin infections, necrotizing fasciitis and toxic shock syndrome, whereas the ST1:USA400 strain results in necrotizing pneumonia and pulmonary sepsis. Other community-acquired strains of MRSA are ST8:USA500 and ST59:USA1000. In many nations of the world, MRSA strains with different predominant genetic background types have come to predominate among CA-MRSA strains; USA300 easily tops the list in the U.S. and is becoming more common in Canada after its first appearance there in 2004. For example, in Australia ST93 strains are common, while in continental Europe ST80 strains, which carry SCC*mec* type IV, predominate. In Taiwan, ST59 strains, some of which are resistant to many non-beta-lactam antibiotics, have arisen as common causes of skin and soft tissue infections in the community. In a remote region of Alaska, unlike most of the continental U.S., USA300 was found rarely in a study of MRSA strains from outbreaks in 1996 and 2000 as well as in surveillance from 2004–06. In 2015, CA-MRSA has been reported in the urban and rural school settings of Kurdistan, Iraq.

In June 2011, the discovery of a new strain of MRSA was announced by two separate teams of researchers in the UK. Its genetic makeup was reportedly more similar to strains found in animals, and testing kits designed to detect MRSA were unable to identify it. This MRSA strain, Clonal Complex 398 (CC398), is responsible for Livestock-associated MRSA (LA-MRSA) infections. Although it is known to be more persistent in colonizing pigs and calves, there have been cases of LA-MRSA carriers with pneumonia, endocarditis, and necrotising fasciitis.

Prevention

Screening Programs

Patient screening upon hospital admission, with nasal cultures, prevents the cohabitation of MRSA carriers with non-carriers, and exposure to infected surfaces. The test used (whether a rapid molecular method or traditional culture) is not as important as the implementation of active screening. In the United States and Canada, the Centers for Disease Control and Prevention issued guidelines on October 19, 2006, citing the need for additional research, but declined to recommend such screening.

In some UK hospitals screening for MRSA is performed in every patient and all NHS surgical patients, except for minor surgeries, are previously checked for MRSA. There is no community screening in the UK; however, screening of individuals is offered by some private companies.

In a US cohort of 1300 healthy children, 2.4% carried MRSA in their nose.

Surface Sanitizing

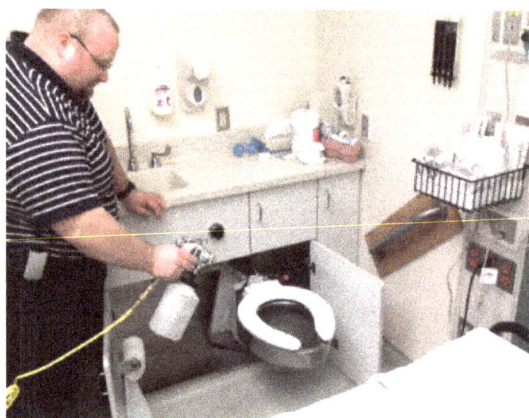

NAV-CO2 sanitizing in Pennsylvania hospital exam room

Alcohol has been proven to be an effective surface sanitizer against MRSA. Quaternary ammonium compounds can be used in conjunction with alcohol to extend the longevity of the sanitizing action. The prevention of nosocomial infections involves routine and terminal cleaning. Non-flammable alcohol vapor in carbon dioxide systems (NAV-CO2) do not corrode metals or plastics used in medical environments and do not contribute to antibacterial resistance.

In healthcare environments, MRSA can survive on surfaces and fabrics, including privacy curtains or garments worn by care providers. Complete surface sanitation is necessary to eliminate MRSA in areas where patients are recovering from invasive procedures. Testing patients for MRSA upon admission, isolating MRSA-positive patients, decolonization of MRSA-positive patients, and terminal cleaning of patients' rooms and all other clinical areas they occupy is the current best practice protocol for nosocomial MRSA.

Studies published from 2004-2007 reported hydrogen peroxide vapor could be used to decontaminate busy hospital rooms, despite taking significantly longer than traditional cleaning. One study noted rapid recontamination by MRSA following the hydrogen peroxide application.

Also tested, in 2006, was a new type of surface cleaner, incorporating accelerated hydrogen peroxide, which was pronounced "a potential candidate" for use against the targeted microorganisms.

Research on Copper Alloys

In 2008, after evaluating a wide body of research mandated specifically by the United States Environmental Protection Agency (EPA), registration approvals were granted by EPA in 2008 granting that copper alloys kill more than 99.9% of MRSA within two hours.

Subsequent research conducted at the University of Southampton (UK) compared the antimicrobial efficacies of copper and several non-copper proprietary coating products to kill MRSA. At 20 °C, the drop-off in MRSA organisms on copper alloy C11000 is dramatic and almost complete (over 99.9% kill rate) within 75 minutes. However, neither a triclosan-based product nor two silver-containing based antimicrobial treatments (Ag-A and Ag-B) exhibited any meaningful efficacy against MRSA. Stainless steel S30400 did not exhibit any antimicrobial efficacy.

In 2004, the University of Southampton research team was the first to clearly demonstrate that copper inhibits MRSA. On copper alloys — C19700 (99% copper), C24000 (80% copper), and C77000 (55% copper) — significant reductions in viability were achieved at room temperatures after 1.5 hours, 3.0 hours and 4.5 hours, respectively. Faster antimicrobial efficacies were associated with higher copper alloy content. Stainless steel did not exhibit any bactericidal benefits.

Hand Washing

In September 2004, after a successful pilot scheme to tackle MRSA, the UK National Health Service announced its *Clean Your Hands* campaign. Wards were required to ensure that alcohol-based hand rubs are placed near all beds so that staff can hand wash more regularly. It is thought that even if this cuts infection by no more than 1%, the plan will pay for itself many times over.

As with some other bacteria, MRSA is acquiring more resistance to some disinfectants and antiseptics. Although alcohol-based rubs remain somewhat effective, a more effective strategy is to wash hands with running water and an antimicrobial cleanser with persistent killing action, such as chlorhexidine. In another study chlorhexidine (Hibiclens), *p*-chloro-*m*-xylenol (Acute-Kare), hexachlorophene (Phisohex), and povidone-iodine (Betadine) were evaluated for their effectiveness. Of the four most commonly used antiseptics, povidone-iodine, when diluted 1:100, was the most rapidly bactericidal against both MRSA and methicillin-susceptible S. *aureus*.

A June 2008 report, centered on a survey by the Association for Professionals in Infection Control and Epidemiology, concluded that poor hygiene habits remain the principal barrier to significant reductions in the spread of MRSA.

Proper Disposal of Hospital Gowns

Used paper hospital gowns are associated with MRSA hospital infections, which could be avoided by proper disposal.

Isolation

Excluding medical facilities, current US guidance does not require workers with MRSA infections to be routinely excluded from the general workplace. Therefore, unless directed by a health care provider, exclusion from work should be reserved for those with wound drainage that cannot be covered and contained with a clean, dry bandage and for those who cannot maintain good hygiene practices. Workers with active infections should be excluded from activities where skin-to-skin contact is likely to occur until their infections are healed. Health care workers should follow the Centers for Disease Control and Prevention's Guidelines for Infection Control in Health Care Personnel.

To prevent the spread of staph or MRSA in the workplace, employers should ensure the availability of adequate facilities and supplies that encourage workers to practice good hygiene; that surface sanitizing in the workplace is followed; and that contaminated equipment are sanitized with Environmental Protection Agency (EPA)-registered disinfectants.

Restricting Antibiotic Use

Glycopeptides, cephalosporins, and, in particular, quinolones are associated with an increased risk of colonisation of MRSA. Reducing use of antibiotic classes that promote MRSA colonisation, especially fluoroquinolones, is recommended in current guidelines.

Public Health Considerations

The burden of MRSA is significant. In 2009, there were an estimated 463,017 (95% confidence interval: 441,595, 484,439) MRSA-related hospitalizations, or a rate of 11.74 (95% confidence interval: 11.20, 12.28) per 1,000 hospitalizations. Many of these infections are less serious, but the Centers for Disease Control and Prevention (CDC) estimates that there are 80,461 invasive MRSA infections and 11,285 deaths due to MRSA annually.

Mathematical models describe one way in which a loss of infection control can occur after measures for screening and isolation seem to be effective for years, as happened in the UK. In the "search and destroy" strategy that was employed by all UK hospitals until the mid-1990s, all patients with MRSA were immediately isolated, and all staff were screened for MRSA and were prevented from working until they had completed a course of eradication therapy that was proven to work. Loss of control occurs because colonised patients are discharged back into the community and then readmitted; when the number of colonised patients in the community reaches a certain threshold, the "search and destroy" strategy is overwhelmed. One of the few countries not to have been overwhelmed by MRSA is the Netherlands: An important part of the success of the Dutch strategy may have been to attempt eradication of carriage upon discharge from hospital.

The Centers for Disease Control and Prevention (CDC) estimated that about 1.7 million nosocomial infections occurred in the United States in 2002, with 99,000 associated deaths. The estimated incidence is 4.5 nosocomial infections per 100 admissions, with direct costs (at 2004 prices) ranging from \$10,500 (£5300, €8000 at 2006 rates) per case (for bloodstream, urinary tract, or respiratory infections in immunocompetent patients) to \$111,000 (£57,000, €85,000) per case for antibiotic-resistant infections in the bloodstream in patients with transplants. With these numbers, conservative estimates of the total direct costs of nosocomial infections are above \$17 billion. The reduction of such infections forms an important component of efforts to improve healthcare safety. (BMJ 2007) MRSA alone was associated with 8% of nosocomial infections reported to the CDC National Healthcare Safety Network from January 2006 to October 2007.

This problem is not unique to one country; the British National Audit Office estimated that the incidence of nosocomial infections in Europe ranges from 4% to 10% of all hospital admissions. As of early 2005, the number of deaths in the United Kingdom attributed to MRSA has been estimated by various sources to lie in the area of 3,000 per year. *Staphylococcus* bacteria account for almost half of all UK hospital infections. The issue of MRSA infections in hospitals has recently been a major political issue in the UK, playing a significant role in the debates over health policy in the United Kingdom general election held in 2005.

On January 6, 2008, half of 64 non-Chinese cases of MRSA infections in Hong Kong in 2007 were Filipino domestic helpers. Ho Pak-leung, professor of microbiology at the University of Hong Kong, traced the cause to high use of antibiotics. In 2007, there were 166 community cases in Hong Kong compared with 8,000 hospital-acquired MRSA cases (155 recorded cases—91 involved Chinese locals, 33 Filipinos, 5 each for Americans and Indians, and 2 each from Nepal, Australia, Denmark and England).

Worldwide, an estimated 2 billion people carry some form of *S. aureus*; of these, up to 53 million (2.7% of carriers) are thought to carry MRSA. In the United States, 95 million carry *S. aureus* in their noses; of these, 2.5 million (2.6% of carriers) carry MRSA. A population review conducted in three U.S. communities showed the annual incidence of CA-MRSA during 2001–2002 to be 18–25.7/100,000; most CA-MRSA isolates were associated with clinically relevant infections, and 23% of patients required hospitalization.

One possible contribution to the increased spread of MRSA infections comes from the use of antibiotics in intensive pig farming. A 2008 study in Canada found MRSA in 10% of tested pork chops and ground pork; a U.S. study in the same year found MRSA in the noses of 70% of the tested farm pigs and in 45% of the tested pig farm workers. There have also been anecdotal reports of increased MRSA infection rates in rural communities with pig farms.

Healthcare facilities with high bed occupancy rates, high levels of temporary nursing staff, or low cleanliness scores no longer have significantly higher MRSA rates. Simple tabular evidence helps provide a clear picture of these changes, showing, for instance, that hospitals with occupancy over 90% had, in 2006–2007, MRSA rates little above those in hospitals with occupancy below 85%, in contrast to the period 2001–2004. In one sense, the disappearance of these relationships is puzzling. Reporters now blame IV cannula and catheters for spreading MRSA in hospitals. (Hospital organisation and speciality mix, 2008)

Decolonization

Care should be taken when trying to drain boils, as disruption of surrounding tissue can lead to larger infections, or even infection of the blood stream (often with fatal consequences). Any drainage should be disposed of very carefully. After the drainage of boils or other treatment for MRSA, patients can shower at home using chlorhexidine (Hibiclens) or hexachlorophene (Phisohex) antiseptic soap (available over-the-counter at many pharmacies) from head to toe. Alternatively, a dilute bleach bath can be taken at a concentration of 2.5 μL/mL dilution of bleach (about 1/2 cup bleach per 1/4-full bathtub of water). Care should be taken to use a clean towel, and to ensure that nasal discharge doesn't infect the towel.

All infectious lesions should be kept covered with a dressing. Mupirocin (Bactroban) 2% ointment can be effective at reducing the size of lesions. A secondary covering of clothing is preferred. As shown in an animal study with diabetic mice, the topical application of a mixture of sugar (70%) and 3% povidone-iodine paste is an effective agent for the treatment of diabetic ulcers with MRSA infection.

The nose is a common refuge for MRSA, and a test swab can be taken of the nose to indicate whether MRSA is present. If MRSA is detected via nasal culture, Mupirocin (Bactroban) 2% ointment can be applied inside each nostril twice daily for 7 days, using a cotton-tipped swab. However, care should be taken so that the swab doesn't penetrate into the sinus. Household members are recommended to follow the same decolonization protocol. After treatment, the nose should be swabbed again to ensure that the treatment was effective. If not, the process should be repeated.

In the hospital setting toilet seats are a common vector for infection, and wiping seats clean before and/or after use can help to prevent the spread of MRSA. Door handles, faucets, light switches, etc. can be disinfected regularly with disinfectant wipes. Spray disinfectants can be used on upholstery. Carpets can be washed with disinfectant, and hardwood floors can be scrubbed with diluted tea tree oil (e.g. Melaleuca). Laundry soap containing tea tree oil may be effective at decontaminating clothing and bedding, especially if hot water and heavy soil cycles are used, however tea tree oil may cause a rash which MRSA can re-colonize. Alcohol-based sanitizers can be placed near bedsides, near sitting areas, in vehicles etc. to encourage their use.

Doctors may also prescribe antibiotics such as clindamycin, doxycycline, or trimethoprim/sulfamethoxazole.

Community Settings

The CDC offers suggestions for preventing the contraction and spread MRSA infection which are applicable to those in community settings, including incarcerated populations, childcare center employees, and athletes. To prevent MRSA infection, individuals should regularly wash hands using soap and water or an alcohol-based sanitizer, keep wounds clean and covered, avoid contact with other people's wounds, avoid sharing personal items such as razors or towels, shower after exercising at athletic facilities (including gyms, weight rooms, and school facilities), shower before using swimming pools or whirlpools, and maintain a clean environment.

It may be difficult for people to maintain the necessary cleanliness if they do not have access to facilities such as public toilets with handwashing facilities. In the United Kingdom, the Workplace (Health, Safety and Welfare) Regulations 1992 requires businesses to provide toilets for their employees, along with washing facilities including soap or other suitable means of cleaning. Guidance on how many toilets to provide and what sort of washing facilities should be provided alongside them is given in the Workplace (Health, Safety and Welfare) Approved Code of Practice and Guidance L24, available from Health and Safety Executive Books. But there is no legal obligation on local authorities in the United Kingdom to provide public toilets, and although in 2008 the House of Commons Communities and Local Government Committee called for a duty on local authorities to develop a public toilet strategy this was rejected by the Government.

Treatment

Both CA-MRSA and HA-MRSA are resistant to traditional anti-staphylococcal beta-lactam antibiotics, such as cephalexin. CA-MRSA has a greater spectrum of antimicrobial susceptibility, including to sulfa drugs (like co-trimoxazole (trimethoprim/sulfamethoxazole)), tetracyclines (like doxycycline and minocycline) and clindamycin (for osteomyelitis), but the drug of choice for treating CA-MRSA is now believed to be vancomycin, according to a Henry Ford Hospital Study. HA-MRSA is resistant even to these antibiotics and often is susceptible only to vancomycin. Newer drugs, such as linezolid (belonging to the newer oxazolidinones class) and daptomycin, are effective against both CA-MRSA and HA-MRSA. The Infectious Disease Society of America recommends vancomycin, linezolid, or clindamycin (if susceptible) for treating patients with MRSA pneumonia. Ceftaroline, a fifth-generation cephalosporin, is the first beta-lactam antibiotic approved in the US to treat MRSA infections (skin and soft tissue or community acquired pneumonia only).

Vancomycin and teicoplanin are glycopeptide antibiotics used to treat MRSA infections. Teicoplanin is a structural congener of vancomycin that has a similar activity spectrum but a longer half-life. Because the oral absorption of vancomycin and teicoplanin is very low, these agents must be administered intravenously to control systemic infections. Treatment of MRSA infection with vancomycin can be complicated, due to its inconvenient route of administration. Moreover, many clinicians believe that the efficacy of vancomycin against MRSA is inferior to that of anti-staphylococcal beta-lactam antibiotics against methicillin-susceptible *Staphylococcus aureus* (MSSA).

Several newly discovered strains of MRSA show antibiotic resistance even to vancomycin and teicoplanin. These new evolutions of the MRSA bacterium have been dubbed vancomycin intermediate-resistant *Staphylococcus aureus* (VISA). Linezolid, quinupristin/dalfopristin, daptomycin, ceftaroline, and tigecycline are used to treat more severe infections that do not respond to glycopeptides such as vancomycin. Current guidelines recommend daptomycin for VISA bloodstream infections and endocarditis.

History

US and UK

In 1959 methicillin was licensed in England to treat penicillin-resistant *S. aureus* infections. Just as bacterial evolution had allowed microbes to develop resistance to penicillin, strains of *S. aureus* evolved to become resistant to methicillin. In 1961 the first known MRSA isolates were reported in a British study, and between 1961-1967 there were infrequent hospital outbreaks in Western Europe and Australia. The first United States hospital outbreak of MRSA occurred at the Boston City Hospital in 1968. Between 1968-mid-1990s the percent of *S. aureus* infections that were caused by MRSA increased steadily, and MRSA became recognized as an endemic pathogen. In 1974 2%

of hospital-acquired *S. aureus* infections could be attributed to MRSA. The rate had increased to 22% by 1995, and by 1997 the percent of hospital *S. aureus* infections attributable to MRSA had reached 50%.

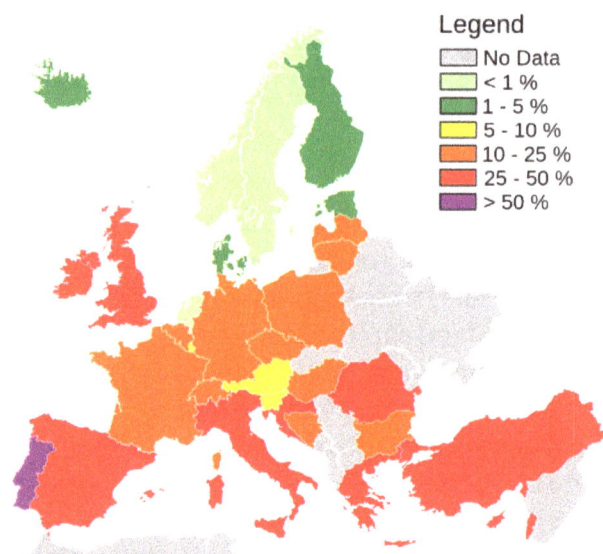

Incidence of MRSA in human blood samples in countries which took part in the study in 2008

The first report of CA-MRSA occurred in 1981, and in 1982 there was a large outbreak of CA-MRSA among intravenous drug users in Detroit, Michigan. Additional outbreaks of CA-MRSA were reported through the 1980s and 1990s, including outbreaks among Australian Aboriginal populations that had never been exposed to hospitals. In the mid-1990s there were scattered reports of CA-MRSA outbreaks among US children. While HA-MRSA rates stabilized between 1998–2008, CA-MRSA rates continued to rise. A report released by the University of Chicago Children's Hospital comparing two time periods (1993–1995 and 1995–1997) found a 25-fold increase in the rate of hospitalizations due to MRSA among children in the United States. In 1999 the University of Chicago reported the first deaths from invasive MRSA among otherwise healthy children in the United States. By 2004 MRSA accounted for 64% of hospital-acquired *S. aureus* infections in the United States.

The Office for National Statistics reported 1,629 MRSA-related deaths in England and Wales during 2005, indicating a MRSA-related mortality rate half the rate of that in the United States for 2005, even though the figures from the British source were explained to be high because of "improved levels of reporting, possibly brought about by the continued high public profile of the disease" during the time of the 2005 United Kingdom General Election. MRSA is thought to have caused 1,652 deaths in 2006 in UK up from 51 in 1993.

It has been argued that the observed increased mortality among MRSA-infected patients may be the result of the increased underlying morbidity of these patients. Several studies, however, including one by Blot and colleagues, that have adjusted for underlying disease still found MRSA bacteremia to have a higher attributable mortality than methicillin-susceptible *S. aureus* (MSSA) bacteremia.

A population-based study of the incidence of MRSA infections in San Francisco during 2004–05 demonstrated that nearly 1 in 300 residents suffered from such an infection in the course of a year and that greater than 85% of these infections occurred outside of the healthcare setting. A 2004 study showed that patients in the United States with *S. aureus* infection had, on average, three times the length of hospital stay (14.3 vs. 4.5 days), incurred three times the total cost ($48,824 vs $14,141), and experienced five times the risk of in-hospital death (11.2% vs 2.3%) than patients without this infection. In a meta-analysis of 31 studies, Cosgrove *et al.*, concluded that MRSA bacteremia is associated with increased mortality as compared with MSSA bacteremia (odds ratio= 1.93; 95% CI = 1.93 ± 0.39). In addition, Wyllie *et al.* report a death rate of 34% within 30 days among patients infected with MRSA, a rate similar to the death rate of 27% seen among MSSA-infected patients.

According to the CDC, the most recent estimates of the incidence of healthcare-associated infections that are attributable to MRSA in the United States indicate a decline in such infection rates. Incidence of MRSA central line-associated blood stream infections as reported by hundreds of intensive care units decreased 50–70% from 2001–2007. A separate system tracking all hospital MRSA bloodstream infections found an overall 34% decrease between 2005–2008.

MRSA is sometimes sub-categorised as community-acquired MRSA (CA-MRSA) or healthcare-associated MRSA (HA-MRSA), although the distinction is complex. Some researchers have defined CA-MRSA by the characteristics of patients whom it infects, while others define it by the genetic characteristics of the bacteria themselves. By 2005, identified CA-MRSA risk factors included athletes, military recruits, incarcerated people, emergency room patients, urban children, HIV-positive individuals, men who have sex with men, and indigenous populations.

By 2015 the proportion of resistant infections in the UK had dropped from 40% down to about 15% a result of intense efforts in hospital hygiene.

Worldwide

The first reported cases of CA-MRSA began to appear in the mid-1990s in Australia, New Zealand, the United States, the United Kingdom, France, Finland, Canada and Samoa, and were notable because they involved people who had not been exposed to a healthcare setting.

Because measurement and reporting varies, it is difficult to compare rates of MRSA in different countries. An international comparison of 2004 MRSA-attributable *S. aureus* rates in middle and high income countries released by the Center For Disease Dynamics, Economics, and Policy in showed that Iceland had the lowest rate of infection, and Romania had the highest at over 70%.

Across Europe in 2015 significant improvements had been made only in Bulgaria, Poland and the British Isles.

Research

Clinical

Many antibiotics against MRSA are in phase II and phase III clinical trials. e.g.:

- Phase III: ceftobiprole, ceftaroline, dalbavancin, telavancin, torezolid, iclaprim, and others.

- Phase II: nemonoxacin.

Development of Aurograb, a treatment intended to complement antibiotics used to treat MRSA, was discontinued after showing a lack of efficacy in Phase II trials.

It has been reported that maggot therapy to clean out necrotic tissue of MRSA infection has been successful. Studies in diabetic patients reported significantly shorter treatment times than those achieved with standard treatments.

Pre-clinical

Phage Therapy

An entirely different approach is phage therapy (e.g., at the George Eliava Institute in Georgia). Experimental phage therapy tested in mice had a reported efficacy against up to 95% of tested *Staphylococcus* isolates.

Antibiotics

- On May 18, 2006, a report in *Nature* identified a new antibiotic, called platensimycin, that had demonstrated successful use against MRSA.

- A new class of non-β-lactam antibiotics, oxadiazoles, was reported to be effective against MRSA infection in mouse models. The mechanisms of oxadiazoles' antibacterial effect are the inhibition of the penicillin binding protein, PBP2a and biosynthesis of the bacterial cell wall. It was found to have bactericidal activity against vancomycin- and linezolid-resistant MRSA and other Gram-positive bacterial strains.

Other Treatments

- Some *in vitro* studies with honey have identified components in honey that kill MRSA.

- Ocean-dwelling living sponges produce compounds that may make MRSA more susceptible to antibiotics.

- Some semi-toxic fungi/mushrooms excrete broad-spectrum antibiotics, not all of which have been fully identified; some have been shown to inhibit the growth of *Staphylococcus aureus*.

- An *in vitro* study showed that the cannabinoids CBD and CBG inhibit MRSA, in addition to the terpenoid pinene which occurs in cannabis.

- Cannabinoids (components of Cannabis sativa), including cannabidiol (CBD), cannabinol (CBN), cannabichromene (CBC), tetrahydrocannabinol (THC), and cannabigerol (CBG), show activity against a variety of MRSA strains.

- *In vitro* studies have shown that oakin, an oak extract, can kill MRSA.

- A 1,000-year-old eye salve recipe found in the medieval Bald's Leechbook at the British Library, one of the earliest known medical textbooks, was found to have activity against MRSA *in vitro* and in skin wounds in mice.

- Some studies suggest that allicin, a compound found in garlic, may prove to be effective in the treatment of MRSA.

Additional Images

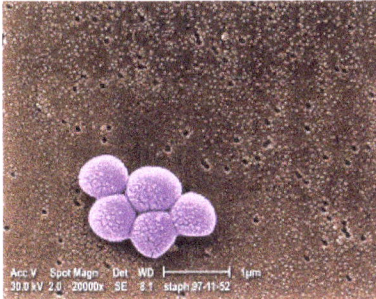

A colourised SEM of MRSA

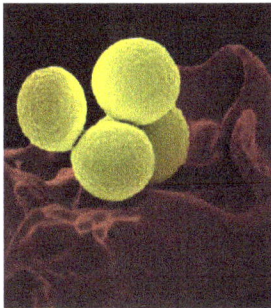

Scanning electron micrograph of a human neutrophil ingesting MRSA

Salmonella

Salmonella is a genus of rod-shaped (bacillus) gram-negative bacterium of the Enterobacteriaceae family. The two species of *Salmonella* are *Salmonella enterica* and *Salmonella bongori*. *Salmonella enterica* is the type species and is further divided into six subspecies that include over 2500 serovars.

S. enterica subspecies are found worldwide in all warm-blooded animals, and in the environment. *S. bongori* is restricted to cold-blooded animals particularly reptiles. Strains of *Salmonella* cause illnesses such as typhoid fever, paratyphoid fever, and food poisoning (salmonellosis).

Traits

Salmonella species are non-spore-forming, predominantly motile enterobacteria with cell diameters between approximately 0.7 and 1.5 μm, lengths from 2 to 5 μm, and peritrichous flagella (flagella that are all around the cell body). They are chemotrophs, obtaining their energy from oxidation and reduction reactions using organic sources. They are also facultative anaerobes, capable of surviving with or without oxygen.

Taxonomy

The genus *Salmonella* is part of the family of Enterobacteriaceae. Its taxonomy has been revised and has the potential to confuse. The genus comprises two species, *Salmonella bongori* and *Salmonella enterica*, the latter of which is divided into six subspecies: *enterica*, *salamae*, *arizonae*, *diarizonae*, *houtenae*, and *indica*. The taxonomic group contains more than 2500 serovars, defined on the basis of the somatic O (lipopolysaccharide) and flagellar H antigens (the Kauffman–White classification). The full name of a serovar is given as, for example, *Salmonella enterica* subsp. *enterica* serovar Typhimurium, but can be abbreviated to *Salmonella* Typhimurium. Further differentiation of strains to assist clinicoepidemiological investigation may be achieved by antibiogram and by supra- or subgenomic techniques such as pulsed-field gel electrophoresis, multilocus sequence typing, and, increasingly, whole genome sequencing. Historically, salmonellae have been clinically categorized as invasive (typhoidal) or noninvasive (nontyphoidal salmonellae) based on host preference and disease manifestations in humans.

History

Salmonella was first visualized in 1880 by Karl Eberth in the Peyer's patches and spleens of typhoid patients. Four years later in 1884 Georg Theodor Gaffky was able to successfully grow the pathogen in pure culture. A year after that, medical research scientist Theobald Smith discovered what would be later known as *Salmonella enterica* (var. Choleraesuis). At the time, Smith was working as a research laboratory assistant in the Veterinary Division of the United States Department of Agriculture. The department was under the administration of Daniel Elmer Salmon, a veterinary pathologist. Initially, *Salmonella* Choleraesuis was thought to be the causative agent of hog cholera, so Salmon and Smith named it "Hog-cholerabacillus". The name *Salmonella* was not used until 1900, when Joseph Leon Lignières proposed that the pathogen discovered by Daniel Salmon's group be called *Salmonella* in his honor.

Detection, Culture and Growth Conditions

Most subspecies of *Salmonella* produce hydrogen sulfide, which can readily be detected by growing them on media containing ferrous sulfate, such as is used in the triple sugar iron test. Most isolates exist in two phases: a motile phase I and a nonmotile phase II. Cultures that are nonmotile upon primary culture may be switched to the motile phase using a Craigie tube or ditch plate.

Salmonella can also be detected and subtyped using multiplex or real-time polymerase chain reactions (PCR) from extracted *Salmonella* DNA.

Mathematical models of *Salmonella* growth kinetics have been developed for chicken, pork, tomatoes, and melons. *Salmonella* reproduce asexually with a cell division interval of 40 minutes.

Salmonella species lead predominantly host-associated lifestyles, but the bacteria were found to be able to persist in a bathroom setting for weeks following contamination, and are frequently isolated from water sources, which act as bacterial reservoirs and may help to facilitate transmission between hosts.

The bacteria are not destroyed by freezing, but UV light and heat accelerate their destruction—they perish after being heated to 55 °C (131 °F) for 90 min, or to 60 °C (140 °F) for 12 min. To protect against *Salmonella* infection, heating food for at least 10 minutes to an internal temperature of 75 °C (167 °F) is recommended.

Salmonella species can be found in the digestive tracts of humans and animals, especially reptiles. *Salmonella* on the skin of reptiles or amphibians can be passed to people who handle the animals. Food and water can also be contaminated with the bacteria if they come in contact with the feces of infected people or animals.

Nomenclature

Initially, each *Salmonella* "species" was named according to clinical considerations, for example *Salmonella typhi-murium* (mouse typhoid fever), *S. cholerae-suis*. After it was recognized that host specificity did not exist for many species, new strains (or serovars, short for serological variants) received species names according to the location at which the new strain was isolated. Later, molecular findings led to the hypothesis that *Salmonella* consisted of only one species, *S. enterica*, and the serovars were classified into six groups, two of which are medically relevant. As this now-formalized nomenclature is not in harmony with the traditional usage familiar to specialists in microbiology and infectologists, the traditional nomenclature is still common. Currently, the two recognized species are *S. enterica*, and *S. bongori*. In 2005, a third species, *Salmonella subterranean*, was proposed, but according to the World Health Organization, the bacterium reported does not belong in the genus *Salmonella*. The six main recognised subspecies are: *enterica* (serotype I), *salamae* (serotype II), *arizonae* (IIIa), *diarizonae* (IIIb), *houtenae* (IV), and *indica* (VI). The former serotype (V) was *bongori*, which is now considered its own species.

The serovar, or serotype, is a classification of *Salmonella* into subspecies based on antigens that the organism presents. It is based on the Kauffman-White classification scheme that differentiates serological varieties from each other. Serotypes are usually put into subspecies groups after the genus and species, with the serovars/serotypes capitalized, but not italicized: An example is *Salmonella enterica* serovar Typhimurium. More modern approaches for typing and subtyping *Salmonella* include DNA-based methods such as pulsed field gel electrophoresis, multiple-loci VNTR analysis, multilocus sequence typing, and multiplex-PCR-based methods.

As Pathogens

Salmonella species are facultative intracellular pathogens. Many infections are due to ingestion of contaminated food. *Salmonella* serovars can be divided into two main groups—typhoidal and nontyphoidal *Salmonella*. Nontyphoidal serovars are more common, and usually cause self-limiting gastrointestinal disease. They can infect a range of animals, and are zoonotic, meaning they can be transferred between humans and other animals. Typhoidal serovars include *Salmonella* Typhi and *Salmonella* Paratyphi A, which are adapted to humans and do not occur in other animals.

Nontyphoidal Salmonella

Infection with nontyphoidal serovars of *Salmonella* generally results in food poisoning. Infection usually occurs when a person ingests foods that contain a high concentration of the bacteria. Infants and young children are much more susceptible to infection, easily achieved by ingesting a small number of bacteria. In infants, infection through inhalation of bacteria-laden dust is possible.

The organisms enter through the digestive tract and must be ingested in large numbers to cause disease in healthy adults. An infection can only begin after living salmonellae (not merely *Salmo-nella*-produced toxins) reach the gastrointestinal tract. Some of the microorganisms are killed in the stomach, while the surviving ones enter the small intestine and multiply in tissues. Gastric acidity is responsible for the destruction of the majority of ingested bacteria, but *Salmonella* has evolved a degree of tolerance to acidic environments that allows a subset of ingested bacteria to survive. Bacterial colonies may also become trapped in mucus produced in the oesophagus. By the end of the incubation period, the nearby host cells are poisoned by endotoxins released from the dead salmonellae. The local response to the endotoxins is enteritis and gastrointestinal disorder.

About 2,000 serotypes of nontyphoidal *Salmonella* are known, which may be responsible for as many as 1.4 million illnesses in the United States each year. People who are at risk for severe illness include infants, elderly, organ-transplant recipients, and the immunocompromised.

Invasive Nontyphoidal Salmonella Disease

While in developed countries, nontyphoidal serovars present mostly as gastrointestinal disease; in sub-Saharan Africa, these serovars can create a major problem in bloodstream infections, and are the most commonly isolated bacteria from the blood of those presenting with fever. Bloodstream infections caused by nontyphoidal salmonellae in Africa were reported in 2012 to have a case fatality rate of 20–25%. Most cases of invasive nontyphoidal salmonella infection iNTS) are caused by *S. typhimurium* or *S. enteritidis*. A new form of *Salmonella typhimurium* (ST313) emerged in the southeast of the African continent 75 years ago, followed by a second wave which came out of central Africa 18 years later. This second wave of iNTS possibly originated in the Congo Basin, and early in the event picked up a gene that made it resistant to the antibiotic chloramphenicol. This created the need to use expensive antimicrobial drugs in areas of Africa that were very poor, making treatment difficult. The increased prevalence of iNTS in sub-Saharan Africa compared to other regions is thought to be due to the large proportion of the African population with some degree of immune suppression or impairment due to the burden of HIV, malaria, and malnutrition, especially in children. The genetic makeup of iNTS is evolving into a more typhoid-like bacterium, able to efficiently spread around the human body. Symptoms are reported to be diverse, including fever, hepatosplenomegaly, and respiratory symptoms, often with an absence of gastrointestinal symptoms.

Typhoidal Salmonella

Typhoid fever is caused by *Salmonella* serotypes which are strictly adapted to humans or higher primates—these include *Salmonella* Typhi, Paratyphi A, Paratyphi B and Paratyphi C. In the systemic form of the disease, salmonellae pass through the lymphatic system of the intestine into the blood of the patients (typhoid form) and are carried to various organs (liver, spleen, kidneys) to form secondary foci (septic form). Endotoxins first act on the vascular and nervous apparatus, resulting in increased permeability and decreased tone of the vessels, upset of thermal regulation, and vomiting and diarrhoea. In severe forms of the disease, enough liquid and electrolytes are lost to upset the water-salt metabolism, decrease the circulating blood volume and arterial pressure, and cause hypovolemic shock. Septic shock may also develop. Shock of mixed character (with signs of both hypovolemic and septic shock) is more common in severe salmonellosis. Oliguria and azotemia may develop in severe cases as a result of renal involvement due to hypoxia and toxemia.

Global Monitoring

In Germany, food poisoning infections must be reported. Between 1990 and 2005, the number of officially recorded cases decreased from about 200,000 to about 50,000 cases. In the United States, about 50,000 cases of *Salmonella* infection are reported each year. A World Health Organization study estimated that 21,650,974 cases of typhoid fever occurred in 2000, 216,510 of which resulted in death, along with 5,412,744 cases of paratyphoid fever.

Molecular Mechanisms of Infection

Mechanisms of infection differ between typhoidal and nontyphoidal serovars, owing to their different targets in the body and the different symptoms that they cause. Both groups must enter by crossing the barrier created by the intestinal cell wall, but once they have passed this barrier, they use different strategies to cause infection.

Nontyphoidal serovars preferentially enter M cells on the intestinal wall by bacterial-mediated endocytosis, a process associated with intestinal inflammation and diarrhoea. They are also able to disrupt tight junctions between the cells of the intestinal wall, impairing the cells' ability to stop the flow of ions, water, and immune cells into and out of the intestine. The combination of the inflammation caused by bacterial-mediated endocytosis and the disruption of tight junctions is thought to contribute significantly to the induction of diarrhoea.

Salmonellae are also able to breach the intestinal barrier via phagocytosis and trafficking by CD18-positive immune cells, which may be a mechanism key to typhoidal *Salmonella* infection. This is thought to be a more stealthy way of passing the intestinal barrier, and may, therefore, contribute to the fact that lower numbers of typhoidal *Salmonella* are required for infection than nontyphoidal *Salmonella*. *Salmonella* cells are able to enter macrophages via macropinocytosis. Typhoidal serovars can use this to achieve dissemination throughout the body via the mononuclear phagocyte system, a network of connective tissue that contains immune cells, and surrounds tissue associated with the immune system throughout the body.

Much of the success of *Salmonella* in causing infection is attributed to two type III secretion systems which function at different times during infection. One is required for the invasion of non-phagocytic cells, colonization of the intestine, and induction of intestinal inflammatory responses and diarrhea. The other is important for survival in macrophages and establishment of systemic disease. These systems contain many genes which must work co-operatively to achieve infection.

The AvrA toxin injected by the SPI1 type III secretion system of *S.* Typhimurium works to inhibit the innate immune system by virtue of its serine/threonine acetyltransferase activity, and requires binding to eukaryotic target cell phytic acid (IP6). This leaves the host more susceptible to infection.

Salmonellosis is known to be able to cause back pain or spondylosis. It can manifest as five clinical patterns: gastrointestinal tract infection, enteric fever, bacteremia, local infection, and the chronic reservoir state. The initial symptoms are nonspecific fever, weakness, myalgia, etc. In the bacteremia state, it can spread to any parts of the body and this induces localized infection or it forms abscesses. The forms of localized *Salmonella* infections are arthritis, urinary tract infection, infection of the central nervous system, bone infection, soft tissue infection, etc. Infection may remain as the latent form for a long time, and when the function of reticular endothelial cells is deteriorated, it may become activated and consequently, it may secondarily induce spreading infection in the bone several months or several years after acute salmonellosis.

Resistance to Oxidative Burst

A hallmark of *Salmonella* pathogenesis is the ability of the bacterium to survive and proliferate within phagocytic cells. Phagocytic cells produce DNA damaging agents such as nitric oxide and oxygen radicals as a defense against pathogens. Thus, *Salmonella* must face attack by molecules that challenge genome integrity. Buchmeier *et al.* showed that mutants of *Salmonella enterica* lacking RecA or RecBC protein function are highly sensitive to oxidative compounds synthesized by macrophages, and furthermore these findings indicate that successful systemic infection by *S. enterica* requires RecA and RecBC mediated recombinational repair of DNA damage.

Host Adaptation

Salmonella enterica, through some of its serovars such as Typhimurium and Enteriditis, shows signs of the ability to infect several different mammalian host species, while other serovars such as Typhi seem to be restricted to only a few hosts. Some of the ways that *Salmonella* serovars have adapted to their hosts include loss of genetic material and mutation. In more complex mammalian species, immune systems, which include pathogen specific immune responses, target serovars of *Salmonella* through binding of antibodies to structures like flagella. Through the loss of the genetic material that codes for a flagellum to form, *Salmonella* can evade a host's immune system. In the study by Kisela *et al.*, more pathogenic serovars of *S. enterica* were found to have certain adhesins in common that have developed out of convergent evolution. This means that, as these strains of *Salmonella* have been exposed to similar conditions such as immune systems, similar structures evolved separately to negate these similar, more advanced defenses in hosts. There are still many questions about the way that *Salmonella* has evolved into so many different types but it has been suggested that *Salmonella* evolved through several phases. As Baumler *et al.* have suggested, Salmonella most likely evolved through horizontal gene transfer, formation of new serovars due to additional pathogenicity islands and through an approximation of its ancestry. So, *Salmonella* could have evolved into its many different serovars through gaining genetic information from different pathogenic bacteria. The presence of several pathogenicity islands in the genome of different serovars has lent credence to this theory.

Genetics

In addition to its importance as a pathogen, *Salmonella enterica* serovar Typhimurium has been instrumental in the development of genetic tools that led to an understanding of fundamental bacterial physiology. These developments were enabled by the discovery of the first generalized transducing phage, P22, in Typhimurium that allowed quick and easy genetic exchange that allowed fine structure genetic analysis. The large number of mutants led to a revision of genetic nomenclature for bacteria. Many of the uses of transposons as genetic tools, including transposon delivery, mutagenesis, construction of chromosome rearrangements, were also developed in Typhimurium. These genetic tools also led to a simple test for carcinogens, the Ames Test.

Escherichia Coli

Escherichia coli is a gram-negative, facultatively anaerobic, rod-shaped bacterium of the genus *Escherichia* that is commonly found in the lower intestine of warm-blooded organisms (endotherms). Most *E. coli* strains are harmless, but some serotypes can cause serious food poisoning in their hosts, and are occasionally responsible for product recalls due to food contamination. The harmless strains are part of the normal flora of the gut, and can benefit their hosts by producing vitamin K_2, and preventing colonization of the intestine with pathogenic bacteria. *E. coli* is expelled into the environment within fecal matter. The bacterium grows massively in fresh fecal matter under aerobic conditions for 3 days, but its numbers decline slowly afterwards.

E. coli and other facultative anaerobes constitute about 0.1% of gut flora, and fecal–oral transmission is the major route through which pathogenic strains of the bacterium cause disease. Cells are able to survive outside the body for a limited amount of time, which makes them potential indicator organisms to test environmental samples for fecal contamination. A growing body of research, though, has examined environmentally persistent *E. coli* which can survive for extended periods outside of a host.

The bacterium can be grown and cultured easily and inexpensively in a laboratory setting, and has been intensively investigated for over 60 years. *E. coli* is a chemoheterotroph whose chemically defined medium must include a source of carbon and energy. Organic growth factors included in chemically defined medium used to grow *E. coli* includes glucose, ammonium phosphate, mono basic, sodium chloride, magnesium sulfate, potassium phosphate, dibasic, and water. The exact chemical composition is known for media that is considered chemically defined medium. *E. coli* is the most widely studied prokaryotic model organism, and an important species in the fields of biotechnology and microbiology, where it has served as the host organism for the majority of work with recombinant DNA. Under favorable conditions, it takes only 20 minutes to reproduce.

Biology and Biochemistry

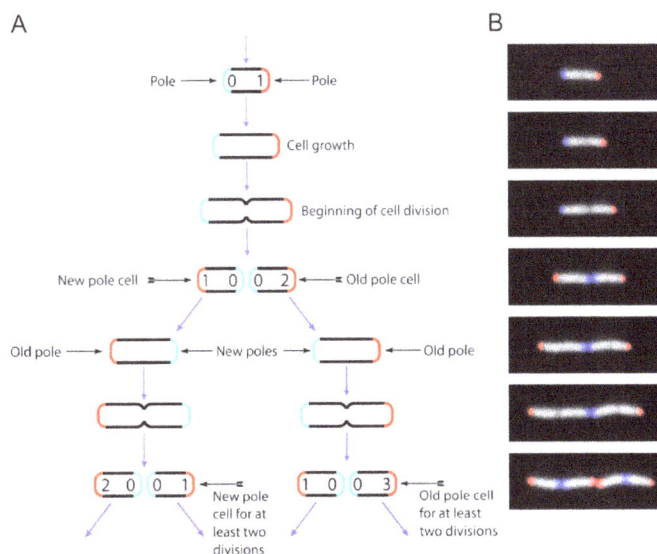

Model of successive binary fission in *E. coli*

A colony of *E. coli* growing

Type and Morphology

E. coli is a gram-negative, facultative anaerobic (that makes ATP by aerobic respiration if oxygen is present, but is capable of switching to fermentation or anaerobic respiration if oxygen is absent) and nonsporulating bacterium. Cells are typically rod-shaped, and are about 2.0 micrometers (μm) long and 0.25–1.0 μm in diameter, with a cell volume of 0.6–0.7 μm^3.

E. coli stains gram-negative because its cell wall is composed of a thin peptidoglycan layer and an outer membrane. During the staining process, *E. coli* picks up the color of the counterstain safranin and stains pink. The outer membrane surrounding the cell wall provides a barrier to certain antibiotics such that *E. coli* is not damaged by penicillin.

Strains that possess flagella are motile. The flagella have a peritrichous arrangement.

Metabolism

E. coli can live on a wide variety of substrates and uses mixed-acid fermentation in anaerobic conditions, producing lactate, succinate, ethanol, acetate, and carbon dioxide. Since many pathways in mixed-acid fermentation produce hydrogen gas, these pathways require the levels of hydrogen to be low, as is the case when *E. coli* lives together with hydrogen-consuming organisms, such as methanogens or sulphate-reducing bacteria.

Culture Growth

Optimum growth of *E. coli* occurs at 37 °C (98.6 °F), but some laboratory strains can multiply at temperatures of up to 49 °C (120.2 °F). Growth can be driven by aerobic or anaerobic respiration, using a large variety of redox pairs, including the oxidation of pyruvic acid, formic acid, hydrogen, and amino acids, and the reduction of substrates such as oxygen, nitrate, fumarate, dimethyl sulfoxide, and trimethylamine N-oxide. *E. coli* is classified as a facultative anaerobe. It uses oxygen when it is present and available. It can however, continue to grow in the absence of oxygen using fermentation or anaerobic respiration. The ability to be able to continue growing in the absence of oxygen is an advantage to bacteria because their survival is increased in environments where water predominates.

Cell Cycle

The bacterial cell cycle is divided into three stages. The B period occurs between the completion of cell division and the beginning of DNA replication. The C period encompasses the time it takes to replicate the chromosomal DNA. The D period refers to the stage between the conclusion of DNA replication and the end of cell division. The doubling rate of *E. coli* is higher when more nutrients are available. However, the length of the C and D periods do not change, even when the doubling time becomes less than the sum of the C and D periods. At the fastest growth rates, replication begins before the previous round of replication has completed, resulting in multiple replication forks along the DNA and overlapping cell cycles.

Unlike eukaryotes, prokaryotes do not rely upon either changes in gene expression or changes in protein synthesis to control the cell cycle. This probably explains why they do not have similar proteins to those used by eukaryotes to control their cell cycle, such as cdk1. This has led to research on what the control mechanism is in prokaryotes. Recent evidence suggests that it may be membrane- or lipid-based.

Genetic Adaptation

E. coli and related bacteria possess the ability to transfer DNA via bacterial conjugation or transduction, which allows genetic material to spread horizontally through an existing population. The process of transduction, which uses the bacterial virus called a bacteriophage, is where the spread of the gene encoding for the Shiga toxin from the *Shigella* bacteria to *E. coli* helped produce *E. coli* O157:H7, the Shiga toxin producing strain of *E. coli*.

Diversity

Escherichia coli encompasses an enormous population of bacteria that exhibit a very high degree of both genetic and phenotypic diversity. Genome sequencing of a large number of isolates of *E. coli* and related bacteria shows that a taxonomic reclassification would be desirable. However, this has not been done, largely due to its medical importance, and *E. coli* remains one of the most diverse bacterial species: only 20% of the genes in a typical *E. coli* genome is shared among all strains.

In fact, from the evolutionary point of view, the members of genus *Shigella* (*S. dysenteriae*, *S. flexneri*, *S. boydii*, and *S. sonnei*) should be classified as *E. coli* strains, a phenomenon termed taxa in disguise. Similarly, other strains of *E. coli* (e.g. the K-12 strain commonly used in recombinant DNA work) are sufficiently different that they would merit reclassification.

A strain is a subgroup within the species that has unique characteristics that distinguish it from other strains. These differences are often detectable only at the molecular level; however, they may result in changes to the physiology or lifecycle of the bacterium. For example, a strain may gain pathogenic capacity, the ability to use a unique carbon source, the ability to take upon a particular ecological niche, or the ability to resist antimicrobial agents. Different strains of *E. coli* are often host-specific, making it possible to determine the source of fecal contamination in environmental samples. For example, knowing which *E. coli* strains are present in a water sample allows researchers to make assumptions about whether the contamination originated from a human, another mammal, or a bird.

Serotypes

A common subdivision system of *E. coli*, but not based on evolutionary relatedness, is by serotype, which is based on major surface antigens (O antigen: part of lipopolysaccharide layer; H: flagellin; K antigen: capsule), e.g. O157:H7). It is, however, common to cite only the serogroup, i.e. the O-antigen. At present, about 190 serogroups are known. The common laboratory strain has a mutation that prevents the formation of an O-antigen and is thus not typeable.

Genome Plasticity and Evolution

Like all lifeforms, new strains of *E. coli* evolve through the natural biological processes of mutation, gene duplication, and horizontal gene transfer; in particular, 18% of the genome of the laboratory strain MG1655 was horizontally acquired since the divergence from *Salmonella*. *E. coli* K-12 and *E. coli* B strains are the most frequently used varieties for laboratory purposes. Some strains develop traits that can be harmful to a host animal. These virulent strains typically cause a bout of diarrhea that is unpleasant in healthy adults and is often lethal to children in the developing world. More virulent strains, such as O157:H7, cause serious illness or death in the elderly, the very young, or the immunocompromised.

The genera *Escherichia* and *Salmonella* diverged around 102 million years ago (credibility interval: 57–176 mya) which coincides with the divergence of their hosts: the former being found in mammals and the latter in birds and reptiles. This was followed by a split of the escherichian ancestor into five species (*E. albertii*, *E. coli*, *E. fergusonii*, *E. hermannii*, and *E. vulneris*). The last *E. coli* ancestor split between 20 and 30 million years ago.

The long-term evolution experiments using *E. coli*, begun by Richard Lenski in 1988, have allowed direct observation of major evolutionary shifts in the laboratory. In this experiment, one population of *E. coli* unexpectedly evolved the ability to aerobically metabolize citrate, which is extremely rare in *E. coli*. As the inability to grow aerobically is normally used as a diagnostic criterion with which to differentiate *E. coli* from other, closely related bacteria, such as *Salmonella*, this innovation may mark a speciation event observed in the laboratory.

Neotype Strain

E. coli is the type species of the genus (*Escherichia*) and in turn *Escherichia* is the type genus of the family Enterobacteriaceae, where the family name does not stem from the genus *Enterobacter* + "i" (sic.) + "aceae", but from "enterobacterium" + "aceae" (enterobacterium being not a genus, but an alternative trivial name to enteric bacterium).

The original strain described by Escherich is believed to be lost, consequently a new type strain (neotype) was chosen as a representative: the neotype strain is U5/41^T, also known under the deposit names DSM 30083, ATCC 11775, and NCTC 9001, which is pathogenic to chickens and has an O1:K1:H7 serotype. However, in most studies, either O157:H7, K-12 MG1655, or K-12 W3110 were used as a representative *E. coli*. The genome of the type strain has only lately been sequenced. Particularly the use of whole genome sequences yields highly supported phylogenies. Based on such data, five subspecies of *E. coli* were distinguished.

The link between phylogenetic distance ("relatedness") and pathology is small, *e.g.* the O157:H7 serotype strains, which form a clade ("an exclusive group")—group E below—are all enterohaemorragic strains (EHEC), but not all EHEC strains are closely related. In fact, four different species of *Shigella* are nested among *E. coli* strains (*vide supra*), while *E. albertii* and *E. fergusonii* are outside of this group. Indeed, all *Shigella* species were placed within a single subspecies of *E. coli* in a phylogenomic study that included the type strain, and for this reason an according reclassification is difficult. All commonly used research strains of *E. coli* belong to group A and are derived mainly from Clifton's K-12 strain (λ^+ F^+; O16) and to a lesser degree from d'Herelle's *Bacillus coli* strain (B strain)(O7).

Genomics

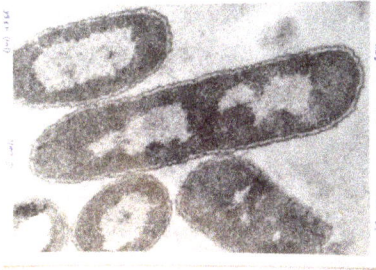

Early electron microscopy

The first complete DNA sequence of an *E. coli* genome (laboratory strain K-12 derivative MG1655) was published in 1997. It was found to be a circular DNA molecule 4.6 million base pairs in length, containing 4288 annotated protein-coding genes (organized into 2584 operons), seven ribosomal RNA (rRNA) operons, and 86 transfer RNA (tRNA) genes. Despite having been the subject of intensive genetic analysis for about 40 years, a large number of these genes were previously unknown. The coding density was found to be very high, with a mean distance between genes of only 118 base pairs. The genome was observed to contain a significant number of transposable genetic elements, repeat elements, cryptic prophages, and bacteriophage remnants.

Today, several hundred complete genomic sequences of *Escherichia* and *Shigella* species are available. The genome sequence of the type strain of *E. coli* has been added to this collection not before 2014. Comparison of these sequences shows a remarkable amount of diversity; only about 20% of each genome represents sequences present in every one of the isolates, while around 80% of each genome can vary among isolates. Each individual genome contains between 4,000 and 5,500 genes, but the total number of different genes among all of the sequenced *E. coli* strains (the pangenome) exceeds 16,000. This very large variety of component genes has been interpreted to mean that two-thirds of the *E. coli* pangenome originated in other species and arrived through the process of horizontal gene transfer.

Gene Nomenclature

Genes in *E. coli* are usually named by 4-letter acronyms that derive from their function (when known). For instance, recA is named after its role in homologous recombination plus the letter A. Functionally related genes are named recB, recC, recD etc. The proteins are named by uppercase acronyms, e.g. RecA, RecB, etc. When the genome of *E. coli* was sequenced, all genes were numbered (more or less) in their order on the genome and abbreviated by b numbers, such as b2819 (=recD) etc. The "b" names were created after Fred Blattner who led the genome sequence effort.

Another numbering system was introduced with the sequence of another *E. coli* strain, W3110, which was sequenced in Japan and hence uses numbers starting by JW... (Japanese W3110), e.g. JW2787 (= recD). Hence, recD = b2819 = JW2787. Note, however, that most databases have their own numbering system, e.g. the EcoGene database uses EG10826 for recD. Finally, ECK numbers are specifically used for alleles in the MG1655 strain of *E. coli* K-12. Complete lists of genes and their synonyms can be obtained from databases such as EcoGene or Uniprot.

Proteomics

Proteome

Several studies have investigated the proteome of *E. coli*. By 2006, 1,627 (38%) of the 4,237 open reading frames (ORFs) had been identified experimentally.

Interactome

The interactome of *E. coli* has been studied by affinity purification and mass spectrometry (AP/MS) and by analyzing the binary interactions among its proteins.

Protein complexes. A 2006 study purified 4,339 proteins from cultures of strain K-12 and found interacting partners for 2,667 proteins, many of which had unknown functions at the time. A 2009 study found 5,993 interactions between proteins of the same *E. coli* strain, though these data showed little overlap with those of the 2006 publication.

Binary interactions. Rajagopala *et al.* (2014) have carried out systematic yeast two-hybrid screens with most *E. coli* proteins, and found a total of 2,234 protein-protein interactions. This study also integrated genetic interactions and protein structures and mapped 458 interactions within 227 protein complexes.

Normal Microbiota

E. coli belongs to a group of bacteria informally known as coliforms that are found in the gastrointestinal tract of warm-blooded animals. *E. coli* normally colonizes an infant's gastrointestinal tract within 40 hours of birth, arriving with food or water or from the individuals handling the child. In the bowel, *E. coli* adheres to the mucus of the large intestine. It is the primary facultative anaerobe of the human gastrointestinal tract. (Facultative anaerobes are organisms that can grow in either the presence or absence of oxygen.) As long as these bacteria do not acquire genetic elements encoding for virulence factors, they remain benign commensals.

Therapeutic Use

Nonpathogenic *E. coli* strain Nissle 1917, also known as Mutaflor, and *E. coli* O83:K24:H31 (known as Colinfant) are used as probiotic agents in medicine, mainly for the treatment of various gastroenterological diseases, including inflammatory bowel disease.

Role in Disease

Most *E. coli* strains do not cause disease, but virulent strains can cause gastroenteritis, urinary tract infections, and neonatal meningitis. It can also be characterized by severe abdominal cramps, diarrhea that typically turns bloody within 24 hours, and sometimes fever. In rarer cases, virulent strains are also responsible for bowel necrosis (tissue death) and perforation without progressing to hemolytic-uremic syndrome, peritonitis, mastitis, septicemia, and gram-negative pneumonia.

There is one strain, *E.coli* #0157:H7, that produces the Shiga toxin (classified as a bioterrorism agent). This toxin causes premature destruction of the red blood cells, which then clog the body's filtering system, the kidneys, causing hemolytic-uremic syndrome (HUS). This in turn causes strokes due to small clots of blood which lodge in capillaries in the brain. This causes the body parts controlled by this region of the brain not to work properly. In addition, this strain causes the buildup of fluid (since the kidneys do not work), leading to edema around the lungs and legs and arms. This increase in fluid buildup especially around the lungs impedes the functioning of the heart, causing an increase in blood pressure.

Uropathogenic *E. coli* (UPEC) is one of the main causes of urinary tract infections. It is part of the normal flora in the gut and can be introduced in many ways. In particular for females, the direction of wiping after defecation (wiping back to front) can lead to fecal contamination of the urogenital orifices. Anal intercourse can also introduce this bacterium into the male urethra, and in switching from anal to vaginal intercourse, the male can also introduce UPEC to the female urogenital system. For more information, see the databases at the end of the article or UPEC pathogenicity.

In May 2011, one *E. coli* strain, O104:H4, was the subject of a bacterial outbreak that began in Germany. Certain strains of *E. coli* are a major cause of foodborne illness. The outbreak started when several people in Germany were infected with enterohemorrhagic *E. coli* (EHEC) bacteria, leading to hemolytic-uremic syndrome (HUS), a medical emergency that requires urgent treatment. The outbreak did not only concern Germany, but also 11 other countries, including regions in North America. On 30 June 2011, the German *Bundesinstitut für Risikobewertung (BfR)* (Federal Institute for Risk Assessment, a federal institute within the German Federal Ministry of Food, Agriculture and Consumer Protection) announced that seeds of fenugreek from Egypt were likely the cause of the EHEC outbreak.

Treatment

The mainstay of treatment is the assessment of dehydration and replacement of fluid and electrolytes. Administration of antibiotics has been shown to shorten the course of illness and duration of excretion of enterotoxigenic *E. coli* (ETEC) in adults in endemic areas and in traveller's diarrhoea, though the rate of resistance to commonly used antibiotics is increasing and they are generally not recommended. The antibiotic used depends upon susceptibility patterns in the particular geographical region. Currently, the antibiotics of choice are fluoroquinolones or azithromycin, with an emerging role for rifaximin. Oral rifaximin, a semisynthetic rifamycin derivative, is an effective and well-tolerated antibacterial for the management of adults with non-invasive traveller's diarrhoea. Rifaximin was significantly more effective than placebo and no less effective than ciprofloxacin in reducing the duration of diarrhoea. While rifaximin is effective in patients with *E. coli*-predominant traveller's diarrhoea, it appears ineffective in patients infected with inflammatory or invasive enteropathogens.

Prevention

ETEC is the type of *E. coli* that most vaccine development efforts are focused on. Antibodies against the LT and major CFs of ETEC provide protection against LT-producing ETEC expressing homologous CFs. Oral inactivated vaccines consisting of toxin antigen and whole cells, i.e. the licensed recombinant cholera B subunit (rCTB)-WC cholera vaccine Dukoral have been developed. There are currently no licensed vaccines for ETEC, though several are in various stages of development. In different trials, the rCTB-WC cholera vaccine provided high (85–100%) short-term protection. An oral ETEC vaccine candidate consisting of rCTB and formalin inactivated *E. coli* bacteria expressing major CFs has been shown in clinical trials to be safe, immunogenic, and effective against severe diarrhoea in American travelers but not against ETEC diarrhoea in young children in Egypt. A modified ETEC vaccine consisting of recombinant *E. coli* strains over expressing the major CFs and a more LT-like hybrid toxoid called LCTBA, are undergoing clinical testing.

Other proven prevention methods for *E. coli* transmission include handwashing and improved sanitation and drinking water, as transmission occurs through fecal contamination of food and water supplies.

Causes and Risk Factors

- Working around livestock
- Consuming unpasteurized dairy product
- Eating undercooked meat
- Drinking impure water

Model Organism in Life Science Research

Role in Biotechnology

Because of its long history of laboratory culture and ease of manipulation, *E. coli* plays an important role in modern biological engineering and industrial microbiology. The work of Stanley Norman Cohen and Herbert Boyer in *E. coli*, using plasmids and restriction enzymes to create recombinant DNA, became a foundation of biotechnology.

E. coli is a very versatile host for the production of heterologous proteins, and various protein expression systems have been developed which allow the production of recombinant proteins in *E. coli*. Researchers can introduce genes into the microbes using plasmids which permit high level expression of protein, and such protein may be mass-produced in industrial fermentation processes. One of the first useful applications of recombinant DNA technology was the manipulation of *E. coli* to produce human insulin.

Many proteins previously thought difficult or impossible to be expressed in *E. coli* in folded form have been successfully expressed in *E. coli*. For example, proteins with multiple disulphide bonds may be produced in the periplasmic space or in the cytoplasm of mutants rendered sufficiently oxidizing to allow disulphide-bonds to form, while proteins requiring post-translational modification such as glycosylation for stability or function have been expressed using the N-linked glycosylation system of *Campylobacter jejuni* engineered into *E. coli*.

Modified *E. coli* cells have been used in vaccine development, bioremediation, production of biofuels, lighting, and production of immobilised enzymes.

Model Organism

E. coli is frequently used as a model organism in microbiology studies. Cultivated strains (e.g. *E. coli* K12) are well-adapted to the laboratory environment, and, unlike wild-type strains, have lost their ability to thrive in the intestine. Many laboratory strains lose their ability to form biofilms. These features protect wild-type strains from antibodies and other chemical attacks, but require a large expenditure of energy and material resources.

In 1946, Joshua Lederberg and Edward Tatum first described the phenomenon known as bacterial conjugation using *E. coli* as a model bacterium, and it remains the primary model to study conjugation. *E. coli* was an integral part of the first experiments to understand phage genetics, and early researchers, such as Seymour Benzer, used *E. coli* and phage T4 to understand the topography of gene structure. Prior to Benzer's research, it was not known whether the gene was a linear structure, or if it had a branching pattern.

E. coli was one of the first organisms to have its genome sequenced; the complete genome of *E. coli* K12 was published by *Science* in 1997.

By evaluating the possible combination of nanotechnologies with landscape ecology, complex habitat landscapes can be generated with details at the nanoscale. On such synthetic ecosystems, evolutionary experiments with *E. coli* have been performed to study the spatial biophysics of adaptation in an island biogeography on-chip.

Studies are also being performed attempting to program *E. coli* to solve complicated mathematics problems, such as the Hamiltonian path problem.

History

In 1885, the German-Austrian pediatrician Theodor Escherich discovered this organism in the feces of healthy individuals. He called it *Bacterium coli commune* because it is found in the colon. Early classifications of prokaryotes placed these in a handful of genera based on their shape and motility (at that time Ernst Haeckel's classification of bacteria in the kingdom Monera was in place).

Bacterium coli was the type species of the now invalid genus *Bacterium* when it was revealed that the former type species ("*Bacterium triloculare*") was missing. Following a revision of *Bacterium*, it was reclassified as *Bacillus coli* by Migula in 1895 and later reclassified in the newly created genus *Escherichia*, named after its original discoverer.

Vancomycin-resistant Enterococcus

Vancomycin-resistant *Enterococcus*, or vancomycin-resistant enterococci (VRE), are bacterial strains of the genus *Enterococcus* that are resistant to the antibiotic vancomycin.

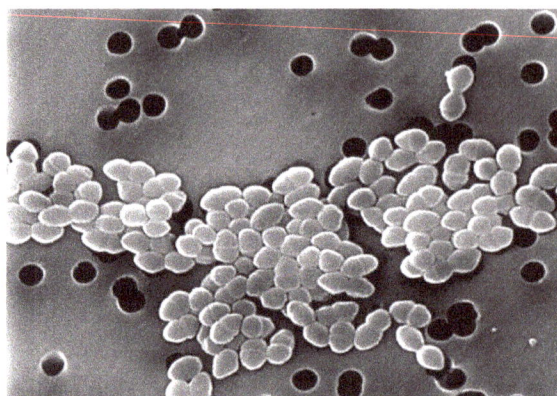

SEM micrograph of vancomycin-resistant enterococci

History and Biology

To become vancomycin-resistant, vancomycin-sensitive enterococci typically obtain new DNA in the form of plasmids or transposons which encode genes that confer vancomycin resistance. This acquired vancomycin resistance is distinguished from the natural vancomycin resistance of certain enterococcal species including *E. gallinarum* and *E. casseliflavus*.

High-level vancomycin-resistant *E. faecalis* and *E. faecium* are clinical isolates first documented in the 1980s. In the United States, vancomycin-resistant *E. faecium* was associated with 4% of healthcare-associated infections reported to the Centers for Disease Control and Prevention National Healthcare Safety Network from January 2006 to October 2007. VRE can be carried by healthy people who have come into contact with the bacteria, usually in a hospital (nosocomial infection), although it is thought that a significant percentage of intensively farmed chicken also carry VRE. Other regions have noted a similar distribution, but with increased incidence of VRE. For example, a 2006 study of nosocomial VRE revealed a rapid spread of resistance among enterococci along with an emerging shift in VRE distribution in the Middle East region, such as Iran. Treatment failures in enterococcal infections result from inadequate information regarding glycopeptide resistance of endemic enterococci due to factors such as the presence of VanA and VanB. The study from Iran reported the first case of VRE isolates that carried VanB gene in enterococcal strains from Iran. This study also noted the first documented isolation of nosocomial *E. raffinosus* and *E. mundtii* in the Middle East region.

Mechanism of Acquired Resistance

Six different types of vancomycin resistance are shown by enterococcus: Van-A, Van-B, Van-C, Van-D, Van-E and Van-G. The significance is that Van-A VRE is resistant to both vancomycin and teicoplanin, Van-B VRE is resistant to vancomycin but susceptible to teicoplanin, and Van-C is only partly resistant to vancomycin, and susceptible to teicoplanin.

The mechanism of resistance to vancomycin found in enterococcus involves the alteration of the peptidoglycan synthesis pathway. The D-alanyl-D-lactate variation results in the loss of one hydrogen-bonding interaction (four, as opposed to five for D-alanyl-D-alanine) being possible between vancomycin and the peptide. The D-alanyl-D-serine variation causes a six-fold loss of affinity between vancomycin and the peptide, likely due to steric hindrance.

Treatment of Infection

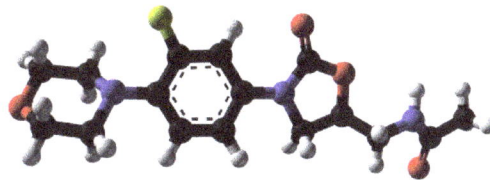

Linezolid

Cephalosporin use is a risk factor for colonization and infection by VRE, and restriction of cephalosporin usage has been associated with decreased VRE infection and transmission in hospitals. *Lactobacillus rhamnosus* GG (LGG), a strain of *L. rhamnosus*, was used successfully for the first time to treat gastrointestinal carriage of VRE. In the US, linezolid is commonly used to treat VRE.

Acinetobacter Baumannii

Acinetobacter baumannii is a typically short, almost round, rod-shaped (coccobacillus) Gram-negative bacterium. It can be an opportunistic pathogen in humans, affecting people with compromised immune systems, and is becoming increasingly important as a hospital-derived (nosocomial) infection. While other species of the genus Acinetobacter are often found in soil samples (leading to the common misconception that *A. baumannii* is a soil organism too), it is almost exclusively isolated from hospital environments. Although occasionally it has been found in environmental soil and water samples, its natural habitat is still not known. Bacteria of this genus lack flagella, whip-like structures many bacteria use for locomotion, but exhibit twitching or swarming motility. This may be due to the activity of type IV pili, pole-like structures that can be extended and retracted. Motility in *A. baumannii* may also be due to the excretion of exopolysaccharide, creating a film of high-molecular-weight sugar chains behind the bacterium to move forward. Clinical microbiologists typically differentiate members of the *Acinetobacter* genus from other Moraxellaceae by performing an oxidase test, as *Acinetobacter* spp. are the only members of the Moraxellaceae to lack cytochrome c oxidases. *A. baumannii* is part of the ACB complex (*A. baumannii*, *A. calcoaceticus*, and *Acinetobacter* genomic species 13TU). Members of the ACB complex are difficult to determine the specific species, and comprise the most clinically relevant members of the genus. *A. baumannii* has also been identified as an ESKAPE pathogen (*Enterococcus faecium*, *Staphylococcus aureus*, *Klebsiella pneumoniae*, *Acinetobacter baumannii*, *Pseudomonas aeruginosa*, and *Enterobacter* species), a group of pathogens with a high rate of antibiotic resistance that are responsible for the majority of nosocomial infections. Colloquially, *A. baumannii* is referred to as 'Iraqibacter' due to its seemingly sudden emergence in military treatment facilities during the Iraq War. It has continued to be an issue for veterans and soldiers who served in Iraq and Afghanistan. Multidrug-resistant *A. baumannii* has spread to civilian hospitals in part due to the transport of infected soldiers through multiple medical facilities.

Virulence Factors and Determinants

Many microbes, including *A. baumannii*, have several properties that allow them to be more successful as pathogens. These properties may be virulence factors such as toxins or toxin delivery systems which directly affect the host cell. They may also be virulence determinants, which are qualities contributing to a microbe's fitness and allow it to survive the host environment, but that do not affect the host directly. These characteristics are just some of the known factors which make *A. baumannii* effective as a pathogen:

AbaR Resistance Islands

Pathogenicity islands, relatively common genetic structures in bacterial pathogens, are composed of two or more adjacent genes that increase a pathogen's virulence. They may contain genes that encode toxins, coagulate blood, or, as in this case, allow the bacteria to resist antibiotics. AbaR-type resistance islands are typical of drug-resistant *A. baumannii*, and different variations may be present in a given strain. Each consists of a transposon backbone of about 16.3 Kb that facilitates horizontal gene transfer. Transposons allow portions of genetic material to be excised from one spot in the genome and integrate into another. This makes horizontal gene transfer of this and similar pathogenicity islands more likely because, when genetic material is taken up by a new bacterium, the transposons allow the pathogenicity island to integrate into the new microorganism's genome. In this case, it would grant the new microorganism the potential to resist certain antibiotics. AbaRs contain several genes for antibiotic resistance, all flanked by insertion sequences. These genes provide resistance to aminoglycosides, aminocyclitols, tetracycline, and chloramphenicol.

Beta-lactamase

A. baumannii has been shown to produce at least one beta-lactamase, which is an enzyme responsible for cleaving the four-atom lactam ring typical of beta-lactam antibiotics. Beta-lactam antibiotics are structurally related to penicillin, which inhibits synthesis of the bacterial cell wall. The cleaving of the lactam ring renders these antibiotics harmless to the bacteria. The beta-lactamase OXA-23 was found to be flanked by insertion sequences, suggesting it was acquired by horizontal gene transfer.

Biofilm Formation

A. baumannii has been noted for its apparent ability to survive on artificial surfaces for an extended period of time, therefore allowing it to persist in the hospital environment. This is thought to be due to its ability to form biofilms. For many biofilm-forming bacteria, the process is mediated by flagella. However, for *A. baumannii,* this process seems to be mediated by pili. Further, disruption of the putative pili chaperone and usher genes *csuC* and *csuE* were shown to inhibit biofilm formation. The formation of biofilms has been shown to alter the metabolism of microorganisms within the biofilm, consequently reducing their sensitivity to antibiotics. This may be because less nutrients are available deeper within the biofilm. A slower metabolism can prevent the bacteria from taking up an antibiotic or performing a vital function fast enough for particular antibiotics to have an effect. They also provide a physical barrier against larger molecules and may prevent desiccation of the bacteria.

Capsule

Many virulent bacteria possess the ability to generate a protective capsule around each individual cell. This capsule, made of long chains of sugars, provides an extra physical barrier between antibiotics, antibodies, and complement. The association of increased virulence with presence of a capsule was classically demonstrated in Griffith's experiment. A gene cluster responsible for secretion of the polysaccharide capsule has been identified and shown to inhibit the antibiotic effect of complement when grown on ascites fluid. A decrease in killing associated with loss of capsule production was then demonstrated using a rat virulence model.

Efflux Pumps

Efflux pumps are protein machines that use energy to pump antibiotics and other small molecules that get into the bacterial cytoplasm and the periplasmic space out of the cell. By constantly pumping antibiotics out of the cell, bacteria can increase the concentration of a given antibiotic required to kill them or inhibit their growth when the target of the antibiotic is inside the bacterium. *A. baumannii* is known to have two major efflux pumps which decrease its susceptibility to antimicrobials. The first, AdeB, has been shown to be responsible for aminoglycoside resistance. The second, AdeDE, is responsible for efflux of a wide range of substrates, including tetracycline, chloramphenicol, and various carbapenems.

Small RNA

Bacterial small RNAs are non-coding RNAs that regulate various cellular processes. Three sRNAs: AbsR11, AbsR25 and AbsR28 have been experimentally validated in the MTCC 1425 (ATCC15308) strain, which is an MDR (multi drug resistance) strain showing resistance to 12 antibiotics. AbsR25 sRNA could play a role in the efflux pump regulation and drug resistance.

OMPA

Adhesion can be a critical determinant of virulence for bacteria. The ability to attach to host cells allows bacteria to interact with them in various ways, whether by type III secretion system or simply by holding on against the prevailing movement of fluids. Outer membrane protein A has been shown to be involved in the adherence of *A. baumannii* to epithelial cells. This allows the bacteria to invade the cells through the zipper mechanism. The protein was also shown to localize to the mitochondria of epithelial cells and cause necrosis by stimulating the production of ROS.

Course of Treatment for Infection

Traumatic injuries, like those from improvised explosive devices, leave large open areas contaminated with debris that are vulnerable to becoming infected with *A. baumannii*.

Because most infections are now resistant to multiple drugs, it is necessary to determine what susceptibilities the particular strain has for treatment to be successful. Traditionally, infections were treated with imipenem or meropenem, but a steady rise in carbapenem-resistant *A. baumannii* has been noted. Consequently, treatment methods often fall back on polymyxins, particularly colistin. Colistin is considered a drug of last resort because it often causes kidney damage among other side effects. Prevention methods in hospitals focus on increased hand-washing and more diligent sterilization procedures.

General Flowchart for Casualty
Transport Between Facilities

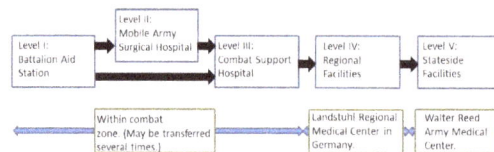

References:
Meghoo, C.A. et al. (2012)
Murray, C.K. (2008)
United States (2009). FM 4-02.1

The logistics of transporting wounded soldiers result in patients visiting several facilities where they may acquire *A. baumannii* infections.

Occurrence in Veterans Injured in Iraq and Afghanistan

Soldiers in Iraq and Afghanistan are at risk for traumatic injury due to gunfire and improvised explosive devices. Previously, infection was thought to occur due to contamination with *A. baumannii* at the time of injury. Subsequent studies have shown, although *A. baumannii* may be infrequently isolated from the natural environment, the infection is more likely nosocomially acquired, likely due to the ability of *A. baumannii* to persist on artificial surfaces for extended periods, and the several facilities to which injured soldiers are exposed during the casualty-evacuation process. Injured soldiers are first taken to level-I facilities, where they are stabilized. Depending on the severity of the injury, the soldiers may then be transferred to a level-II facility, which consists of a forward surgical team, for additional stabilization. Depending on the logistics of the locality, the injured soldiers may transfer between these facilities several times before finally being taken to a major hospital within the combat zone (level III). Generally after 1–3 days, when the patients are stabilized, they are transferred by air to a regional facility (level IV) for additional treatment. For soldiers serving in Iraq or Afghanistan, this is typically Landstuhl Regional Medical Center in Germany. Finally, the injured soldiers are transferred to hospitals in their home country for rehabilitation and additional treatment. This repeated exposure to many different medical environments seems to be the reason *A. baumannii* infections have become increasingly common. Multidrug-resistant *A. baumannii* is a major factor in complicating the treatment and rehabilitation of injured soldiers, and has led to additional deaths.

Incidence of A. baumannii in Hospitals

The importation of *A. baumannii* and subsequent presence in hospitals has been well documented. *A. baumannii* is usually introduced into a hospital by a colonized patient. Due to its ability to survive on artificial surfaces and resist desiccation, it can remain and possibly infect new patients for some time. *A baumannii* growth is suspected to be favored in hospital settings due to the constant use of antibiotics by patients in the hospital. Acinetobacter can be spread by person-to-person contact or contact with contaminated surfaces. In a study of European intensive care units in 2009, *A. baumannii* was found to be responsible for 19.1% of ventilator-associated pneumonia (VAP) cases.

Documented case studies
Country
Australia
Brazil
China
Germany
India
South Korea
United Kingdom
United States

Campylobacter

Campylobacter (meaning "curved bacteria") is a genus of Gram-negative, microaerophilic, oxidase-positive, catalase-positive, nonfermentative bacteria. *Campylobacter* species are typically comma or s-shaped and able to move via unipolar or bipolar flagella. Most *Campylobacter* species are cause for disease and can infect humans and other animals. The bacterium's main reservoir is poultry; humans can contract the disease from contaminated food. Another source of infection is contact with infected animals such as kittens and puppies; most colonized animals -including chickens- develop a lifelong carrier state. At least a dozen species of *Campylobacter* have been implicated in human disease, with *C. jejuni* and *C. coli* being the most common. *C. jejuni* is now recognized as one of the main causes of bacterial foodborne disease in many developed countries. *C. jejuni* infection can also result in serious bacteremia in individuals with AIDS, while *C. lari* is a known cause of recurrent diarrhea in children. *C. fetus* is a cause of spontaneous abortions in cattle and sheep, as well as an opportunistic pathogen in humans.

History

The symptoms of *Campylobacter* infections were described in 1886 in infants by Theodor Escherich. These infections were named cholera infantum, or summer complaint. The genus was first described in 1963; however, the organism was not isolated until 1972.

Genome and Proteome

The genomes of several *Campylobacter* species have been sequenced. The first *Campylobacter* genome to be sequenced was *C. jejuni*, in 2000.

Campylobacter species contain two flagellin genes in tandem for motility, *flaA* and *flaB*. These genes undergo intergenic recombination, further contributing to their virulence. Nonmotile mutants do not colonize.

Sequence features: Comparative genomic analysis has led to the identification of 15 proteins which are uniquely found in members of the genus *Campylobacter* and serve as molecular markers for the genus. Eighteen other proteins were also found which were present in all species except *C. fetus*, which is the deepest-branching *Campylobacter* species. A conserved insertion has also been identified which is present in all *Campylobacter* species except *C. fetus*. Additionally, 28 proteins have been identified present only in *C. jejuni* and *C. coli*, indicating a close relationship between these two species. Five other proteins have also been identified which are only found in *C. jejuni* and serve as molecular markers for the species.

Bacteriophage

The confusing taxonomy of *Campylobacter* over the past decades make it difficult to identify the earliest reports of *Campylobacter* bacteriophages. Bacteriophages specific to the species now known as *C. coli* and *C. fetus* (previously *Vibrio coli* and *V. fetus*), were isolated from cattle and pigs during the 1960s.

Pathogenesis

Campylobacteriosis, a gastrointestinal infection caused by *Campylobacter*, is characterized by inflammatory, sometimes bloody diarrhea or dysentery syndrome, mostly including cramps, fever, and pain. The most common routes of transmission are fecal-oral, ingestion of contaminated food or water, and the eating of raw meat. Foods implicated in campylobacteriosis include raw or under-cooked poultry, raw dairy products, and contaminated produce. *Campylobacter* is sensitive to the stomach's normal production of hydrochloric acid: as a result, the infectious dose is relatively high, and the bacteria rarely cause illness when a person is exposed to less 10,000 organisms. Nevertheless, people taking antacid medication (e. g. people with gastritis or stomach ulcers) are at higher risk of contracting disease from a smaller amount of organisms, since this type of medication inhibits normal gastric acid. The infection is usually self-limiting and, in most cases, symptomatic treatment by liquid and electrolyte replacement is enough in human infections. The use of antibiotics, though, is controversial. Symptoms typically last five to seven days.

The sites of tissue injury include the jejunum, the ileum, and the colon. Most strains of *C jejuni* produce a toxin (cytolethal distending toxin) that hinders the cells from dividing and activating the immune system. This helps the bacteria to evade the immune system and survive for a limited time in the cells. A cholera-like enterotoxin was once thought to be also made, but this appears not to be the case. The organism produces diffuse, bloody, edematous, and exudative enteritis. Although rarely has the infection been considered a cause of hemolytic uremic syndrome and thrombotic thrombocytopenic purpura, no unequivocal case reports exist. In some cases, a *Campylobacter* infection can be the underlying cause of Guillain–Barré syndrome. Gastrointestinal perforation is a rare complication of ileal infection.

Campylobacter has also been associated with periodontitis.

Treatment

Diagnosis of the illness is made by testing a specimen of feces.

- Standard treatment is now azithromycin, a macrolide antibiotic, especially for *Campylobacter* infections in children, although other antibiotics, such as macrolides, quinolones, and tetracycline are sometimes used to treat gastrointestinal *Campylobacter* infections in adults. In case of systemic infections, other bactericidal antibiotics are used, such as ampicillin, amoxicillin/clavulanic acid, or aminoglycosides. Fluoroquinolone antibiotics, such as ciprofloxacin or levofloxacin, may no longer be effective in some cases due to resistance.

- Dehydrated children may require intravenous fluid treatment in a hospital.

- The illness is contagious, and children must be kept at home until they have been clear of symptoms for at least two days.

- Good hygiene is important to avoid contracting the illness or spreading it to others.

- Intestinal perforation is very rare; increased abdominal pain and collapse require immediate medical attention.

Epidemiology

Campylobacter infections increased 14% in the United States in 2012 compared to the period from 2006 to 2008. This represents the highest reported number of infections since 2000.

UK

In January 2013, the UK's Food Standards Agency warned that two-thirds of all raw chicken bought from UK shops was contaminated with *Campylobacter*, affecting an estimated half a million people annually and killing about 100. In June 2014, the Food Standards Agency started a campaign against washing raw chicken, as washing can spread germs by splashing. In May 2015, cumulative results for samples taken from fresh chickens between February 2014 and February 2015 were published as official statistics by the FSA, including results presented by major retailers.

The results for the full year show:

- 19% of chickens tested positive for *Campylobacter* within the highest band of contamination.

- 73% of chickens tested positive for the presence of *Campylobacter*.

- 0.1% (five samples) of packaging tested positive at the highest band of contamination.

- 7% of packaging tested positive for the presence of *Campylobacter*.

USA

Larger prevalence of *Campylobacter* (40% or more) has been reported in raw chicken meat in retail stores in the USA. The reported prevalence in retail chicken meat is higher than the reported prevalence by the microbiology performance standard testing collected by the U. S. Department of Agriculture, and the last quarterly progress report on *Salmonella* and *Campylobacter* testing of meat and poultry for July–September 2014, published by the Food Safety and Inspection Service of the U. S. Department of Agriculture, shows a low prevalence of *Campylobacter* spp. in ground chicken meat, but a larger prevalence (20%) in mechanically separated chicken meat (which is sold only for further processing).

Canada

FoodNet Canada conducts surveillance at 3 different sites on behalf of Public Health Agency of Canada. In the 2014 Short Report of FoodNet Canada, they reported:

- "In 2014, Campylobacter and Salmonella remained the most common causes of human enteric illness in the sentinel sites" [3 sites for 2014],

- "Campylobacter was the most prevalent pathogen found on skinless chicken breast in all sites with close to one-half of all samples testing positive.";

- "In turkey in the BC site, Campylobacter was again the most common pathogen found in 2014, as in 2013. Campylobacter was also commonly found in beef and dairy manure samples in the ON site, as in previous years. Campylobacter prevalence in broiler chickens was variable across the sites, ranging from 8.7% - 22%.";

- "...the 2014 FoodNet Canada sampling year have demonstrated that retail meat products, particularly chicken products, remain an important source of human enteric pathogens."

New Zealand

In August 2016, an estimated 4,000+ residents of Havelock North, a town with 13,000 or so residents, had gastric illness after the water supply was thought to be contaminated by campylobacter.

Streptococcus Pyogenes

Streptococcus pyogenes is a species of bacteria. Like most other streptococci, it is clinically important in human illness. It is an infrequent, but usually pathogenic, part of the skin flora. It is the sole species of Lancefield group A and is often called group A streptococcus (GAS), because it displays streptococcal group A antigen on its cell wall. Group A streptococcal infection can cause illness, which typically produces small zones of beta-hemolysis, a complete destruction of red blood cells. (A zone size of 2–3 mm is typical). It is thus also called group A (beta-hemolytic) streptococcus (GABHS).

Like other cocci, streptococci are round bacteria. The name is derived from the Greek meaning pus(pyo)-forming(genes) chain(Strepto) of berries (coccus), because streptococcal cells tend to link together in chains of round cells and a number of infections caused by it produce pus. Streptococci are catalase-negative and Gram-positive. *S. pyogenes* can be cultured on blood agar plates. Under ideal conditions, it has an incubation period of 1 to 3 days.

An estimated 700 million GAS infections occur worldwide each year. While the overall mortality rate for these infections is 0.1%, over 650,000 of the cases are severe and invasive, and have a mortality rate of 25%. Early recognition and treatment are critical; diagnostic failure can result in sepsis and death.

Serotyping

In 1928, Rebecca Lancefield published a method for serotyping *S. pyogenes* based on its M protein, a virulence factor displayed on its surface. Later, in 1946, Lancefield described the serologic classification of *S. pyogenes* isolates based on their surface T-antigen. Four of the 20 T-antigens have been revealed to be pili, which are used by bacteria to attach to host cells. Over 220 M serotypes and about 20 T serotypes are known.

Pathogenesis

S. pyogenes is the cause of many important human diseases, ranging from mild superficial skin infections to life-threatening systemic diseases. Infections typically begin in the throat or skin. The most striking sign is a strawberry-like rash. Examples of mild *S. pyogenes* infections include pharyngitis (strep throat) and localized skin infection (impetigo). Erysipelas and cellulitis are characterized by multiplication and lateral spread of *S. pyogenes* in deep layers of the skin. *S. pyogenes* invasion and multiplication in the fascia can lead to necrotizing fasciitis, a life-threatening condition requiring surgery.

Infections due to certain strains of *S. pyogenes* can be associated with the release of bacterial toxins. Throat infections associated with release of certain toxins lead to scarlet fever. Other toxigenic *S. pyogenes* infections may lead to streptococcal toxic shock syndrome, which can be life-threatening.

S. pyogenes can also cause disease in the form of postinfectious "nonpyogenic" (not associated with local bacterial multiplication and pus formation) syndromes. These autoimmune-mediated complications follow a small percentage of infections and include rheumatic fever and acute postinfectious glomerulonephritis. Both conditions appear several weeks following the initial streptococcal infection. Rheumatic fever is characterised by inflammation of the joints and/or heart following an episode of streptococcal pharyngitis. Acute glomerulonephritis, inflammation of the renal glomerulus, can follow streptococcal pharyngitis or skin infection.

This bacterium remains acutely sensitive to penicillin. Failure of treatment with penicillin is generally attributed to other local commensal organisms producing β-lactamase, or failure to achieve adequate tissue levels in the pharynx. Certain strains have developed resistance to macrolides, tetracyclines, and clindamycin.

Virulence Factors

S. pyogenes has several virulence factors that enable it to attach to host tissues, evade the immune response, and spread by penetrating host tissue layers. A carbohydrate-based bacterial capsule composed of hyaluronic acid surrounds the bacterium, protecting it from phagocytosis by neutrophils. In addition, the capsule and several factors embedded in the cell wall, including M protein, lipoteichoic acid, and protein F (SfbI) facilitate attachment to various host cells. M protein also inhibits opsonization by the alternative complement pathway by binding to host complement regulators. The M protein found on some serotypes is also able to prevent opsonization by binding to fibrinogen. However, the M protein is also the weakest point in this pathogen's defense, as antibodies produced by the immune system against M protein target the bacteria for engulfment by phagocytes. M proteins are unique to each strain, and identification can be used clinically to confirm the strain causing an infection.

Lysogeny

All strains of *S. pyogenes* are polylysogenized, in that they carry one or more bacteriophage on their genomes. Some of the phages may be defective, but in some cases active phage may compensate for defects in others. In general, the genome of *S. pyogenes* strains isolated during disease are >90% identical, they differ by the phage they carry.

Name	Description
Streptolysin O	An exotoxin, one of the bases of the organism's beta-hemolytic property, streptolysin O causes an immune response and detection of antibodies to it; antistreptolysin O (ASO) can be clinically used to confirm a recent infection.
Streptolysin S	A cardiotoxic exotoxin, another beta-hemolytic component, not immunogenic and O_2 stable: A potent cell poison affecting many types of cell including neutrophils, platelets, and subcellular organelles.
Streptococcal pyrogenic exotoxin A (SpeA)	Superantigens secreted by many strains of S. pyogenes: This pyrogenic exotoxin is responsible for the rash of scarlet fever and many of the symptoms of streptococcal toxic shock syndrome, also known as toxic shock like syndrome(TSLS).
Streptococcal pyrogenic exotoxin C (SpeC)	
Streptokinase	Enzymatically activates plasminogen, a proteolytic enzyme, into plasmin, which in turn digests fibrin and other proteins
Hyaluronidase	Hyaluronidase is widely assumed to facilitate the spread of the bacteria through tissues by breaking down hyaluronic acid, an important component of connective tissue. However, very few isolates of S. pyogenes are capable of secreting active hyaluronidase due to mutations in the gene that encode the enzyme. Moreover, the few isolates capable of secreting hyaluronidase do not appear to need it to spread through tissues or to cause skin lesions. Thus, the true role of hyaluronidase in pathogenesis, if any, remains unknown.
Streptodornase	Most strains of S. pyogenes secrete up to four different DNases, which are sometimes called streptodornase. The DNases protect the bacteria from being trapped in neutrophil extracellular traps (NETs) by digesting the NETs' web of DNA, to which are bound neutrophil serine proteases that can kill the bacteria.
C5a peptidase	C5a peptidase cleaves a potent neutrophil chemotaxin called C5a, which is produced by the complement system. C5a peptidase is necessary to minimize the influx of neutrophils early in infection as the bacteria are attempting to colonize the host's tissue. C5a peptidase, although required to degrade the neutrophil chemotaxin C5a in the early stages of infection, is not required for S. pyogenes to prevent the influx of neutrophils as the bacteria spread through the fascia.
Streptococcal chemokine protease	The affected tissue of patients with severe cases of necrotizing fasciitis are devoid of neutrophils. The serine protease ScpC, which is released by S. pyogenes, is responsible for preventing the migration of neutrophils to the spreading infection. ScpC degrades the chemokine IL-8, which would otherwise attract neutrophils to the site of infection.

Diagnosis

Usually, a throat swab is taken to the laboratory for testing. A Gram stain is performed to show Gram-positive cocci in chains. Then, the organism is cultured on blood agar with an added bacitracin antibiotic disk to show beta-hemolytic colonies and sensitivity (zone of inhibition around the disk) for the antibiotic. Culture on agar not containing blood, and then performing the catalase test should show a negative reaction for all streptococci. *S. pyogenes* is CAMP and hippurate tests negative. Serological identification of the organism involves testing for the presence of group-A-specific polysaccharide in the bacterium's cell wall using the Phadebact test.

The rapid pyrrolidonyl arylamidase (PYR) test is used for the presumptive identification of group A beta-hemolytic streptococci. GBS gives a negative finding on this test.

Treatment

The treatment of choice is penicillin, and the duration of treatment is well established as being 10 days minimum. For toxic shock syndrome and necrotizing fasciitis, high-dose penicillin and clindamycin are used. Additionally, for necrotizing fasciitis, surgery is often needed to remove damaged tissue and stop the spread of the infection.

No instance of penicillin resistance has been reported to date, although since 1985, many reports of penicillin tolerance have been made. The reason for the failure of penicillin to treat *S. pyogenes* is most commonly patient noncompliance, but in cases where patients have been compliant with their antibiotic regimen, and treatment failure still occurs, another course of antibiotic treatment with cephalosporins is common.

In individuals with a penicillin allergy, erythromycin, other macrolides, and cephalosporins have been shown to be effective treatments.

Prevention

S. pyogenes infections are best prevented through effective hand hygiene. No vaccines are currently available to protect against *S. pyogenes* infection, although research has been conducted into the development of one. Difficulties in developing a vaccine include the wide variety of strains of *S. pyogenes* present in the environment and the large amount of time and number of people that will be needed for appropriate trials for safety and efficacy of the vaccine.

Applications

Bionanotechnology

Many *S. pyogenes* proteins have unique properties, which have been harnessed in recent years to produce a highly specific "superglue" and a route to enhance the effectiveness of antibody therapy.

Genome Editing

The CRISPR system from this organism that is used to recognize and destroy DNA from invading viruses, stopping the infection was appropriated in 2012 for use as a genome-editing tool that could potentially alter any piece of DNA and later RNA.

Genome

The genome of different strains were sequenced (genome size is 1.8–1.9 Mbp) encoding about 1700-1900 proteins (1700 in strain NZ131, 1865 in strain MGAS5005).

References

- Casjens S. In: Mahy BWJ and Van Regenmortel MHV. Desk Encyclopedia of General Virology. Boston: Academic Press; 2010. ISBN 0-12-375146-2. p. 167.

- Fenner F.. In: Mahy B. W. J. and Van Regenmortal M. H. V.. Desk Encyclopedia of General Virology. 1 ed. Oxford, UK: Academic Press; 2009. ISBN 0-12-375146-2. p. 15.

- Mahy WJ & Van Regenmortel MHV (eds). Desk Encyclopedia of General Virology. Oxford: Academic Press; 2009. ISBN 0-12-375146-2. p. 26.

- Belay ED and Schonberger LB. Desk Encyclopedia of Human and Medical Virology. Boston: Academic Press; 2009. ISBN 0-12-375147-0. p. 497–504.

- Saunders, Venetia A.; Carter, John. Virology: principles and applications. Chichester: John Wiley & Sons; 2007. ISBN 0-470-02387-2. p. 72.

- King AMQ, Lefkowitz E, Adams MJ, Carstens EB. Virus Taxonomy: Ninth Report of the International Committee on Taxonomy of Viruses. Elsevier; 2011. ISBN 0-12-384684-6. p. 6.

- Molecular Biology of the Cell; Fourth Edition. New York and London: Garland Science; 2002 [Retrieved 2014-12-19]. ISBN 0-8153-3218-1.

- Lomonossoff, GP. Recent Advances in Plant Virology. Caister Academic Press; 2011. ISBN 978-1-904455-75-2. Virus Particles and the Uses of Such Particles in Bio- and Nanotechnology.

- Dusenbery, David B. (2009). Living at Micro Scale, pp. 20–25. Harvard University Press, Cambridge, Mass. ISBN 978-0-674-03116-6.

- Hecker M, Völker U (2001). "General stress response of Bacillus subtilis and other bacteria". Adv Microb Physiol. Advances in Microbial Physiology. 44: 35–91. doi:10.1016/S0065-2911(01)44011-2. ISBN 978-0-12-027744-5.

- Dusenbery, David B. (2009). Living at Micro Scale, p. 136. Harvard University Press, Cambridge, Mass. ISBN 978-0-674-03116-6.

- Francois P, Schrenzel J (2008). "Rapid Diagnosis and Typing of Staphylococcus aureus". Staphylococcus: Molecular Genetics. Caister Academic Press. ISBN 978-1-904455-29-5.

- Mackay I M (editor). (2007). Real-Time PCR in Microbiology: From Diagnosis to Characterization. Caister Academic Press. ISBN 978-1-904455-18-9.

- editors; Gillespie, Stephen H.; Hawkey, Peter M. (2006). Principles and practice of clinical bacteriology (2nd ed.). Hoboken, NJ: John Wiley & Sons. ISBN 9780470017968.

- Heymann, Danielle A. Brands; Alcamo, I. Edward; Heymann, David L. (2006). Salmonella. Philadelphia: Chelsea House Publishers. ISBN 0-7910-8500-7. Retrieved 31 July 2015.

- Tortora, Gerard (2010). Microbiology: An Introduction. San Francisco, CA: Benjamin Cummings. pp. 85–87, 161, 165,. ISBN 0-321-55007-2.

- Krieg, N. R.; Holt, J. G., eds. (1984). Bergey's Manual of Systematic Bacteriology. 1 (First ed.). Baltimore: The Williams & Wilkins Co. pp. 408–420. ISBN 0-683-04108-8.

- Farrar J, Hotez P, Junghanss T, Kang G, Lalloo D, White NJ, eds. (2013). Manson's Tropical Diseases (23rd ed.). Oxford: Elsevier/Saunders. ISBN 9780702053061.

- Gould, Dinah; Brooker, Christine (2008-08-20). Infection Prevention and Control: Applied Microbiology for Healthcare. Palgrave Macmillan. ISBN 9781137045928.

- Robinson, D. Ashley; Feil, Edward J.; Falush, Daniel (2010-03-16). Bacterial Population Genetics in Infectious Disease. John Wiley & Sons. ISBN 9780470600115.

- Fong, I. W.; Drlica, Karl (2007-11-15). Antimicrobial Resistance and Implications for the 21st Century. Springer Science & Business Media. ISBN 9780387724188.

- Phoenix, David A.; Harris, Frederick; Dennison, Sarah R. (2014-08-25). Novel Antimicrobial Agents and Strategies. John Wiley & Sons. ISBN 9783527676156.

- Garrity, edited by G. (2000). Bergey's Manual of Systematic Bacteriology Vol. 2, Pts. A & B: The Proteobacteria. (2nd ed., rev. ed.). New York: Springer. p. 454. ISBN 0-387-95040-0.

- Ryan, Kenneth James; Ray, C. George, eds. (2004). Sherris Medical Microbiology: An Introduction to Infectious Diseases (4th ed.). McGraw Hill. pp. 378–80. ISBN 978-0-8385-8529-0.

5

Methods Related to Antimicrobial Resistance

Agar dilution is used to test the efficiency of antibiotics and the technique used to test the vulnerability of bacteria to antibiotics is known as broth microdilution. They are cost effective methods that are available for microbial eradication. This section discusses the methods of antimicrobial resistance detection and analysis in a critical manner providing key analysis to the subject matter.

Agar Dilution

Agar dilution is a method used by researchers to determine the resistance of pathogens to antibiotics. It is the dilution method most frequently used to test the effectiveness of new antibiotics.

Process

The antibiotic to be tested is added to agar, which is then placed in dilution plates and diluted with varying levels of water. After this, the pathogen to be tested is added to each plate, plus a control plate that does not receive any antibiotics. The dilution plates are then incubated at a temperature of 37 degrees Celsius. The plates are then incubated for sixteen to eighteen hours, although incubation time may be less for bacteria populations that divide quickly. After incubation, the plates are examined to determine if bacterial expansion has occurred. The lowest concentration of antibiotics that stopped the spread of the bacteria is considered to be the minimum inhibitory concentration of that bacteria.

Advantages

Agar dilution is considered to be the gold standard of susceptibility testing, or the most accurate way to measure the resistance of bacteria to antibiotics. The results of agar dilution are easily reproduced and they can be monitored at a much cheaper cost than what is required of other dilution methods. Additionally, up to thirty pathogen samples (plus two controls) can be tested at once, so agar dilution is useful for batch tests.

Disadvantages

Each dilution plate in agar testing has to be manually infected with the pathogen to be tested, so agar dilution testing is both labor-intensive and expensive. Unlike broth microdilution tests, agar dilution cannot be used to test more than one antibiotic at a time.

Broth Microdilution

Broth microdilution is a method used to test the susceptibility of bacteria to antibiotics. It is the most commonly used method to perform this test in the United States.

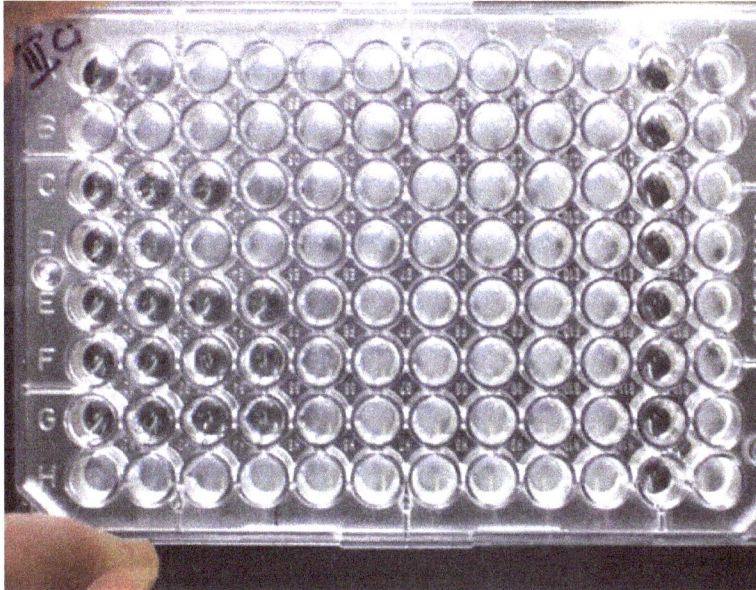

A completed broth microdilution test

Process

During testing, multiple microtiter plates are filled with a broth composed of Brucella and supplements of blood. Varying concentrations of the antibiotics and the bacteria to be tested are then added to the plate. The plate is then placed into a non-CO2 incubator and heated at thirty-five degrees Celsius for sixteen to twenty hours. Following the allotted time, the plate is removed and checked for bacterial growth. If the broth became cloudy or a layer of cells formed at the bottom, then bacterial growth has occurred. The results of the broth microdilution method are reported in Minimum Inhibitory Concentration (MIC), or the lowest concentration of antibiotics that stopped bacterial expansion.

Advantages

The broth microdilution method can be used to test the susceptibility of bacteria to multiple antibiotics at once. Broth microdilution is also highly accurate. The accuracy of its results are comparable to agar dilution, the gold standard of susceptibility testing. Other advantages include the commercial availability of plates, the ease of testing and storing the plates, and the ability for the results of some tests to be read by machines.

References

- Lorian, Victor (2005). Antibiotics in Laboratory Medicine. Lippincott Williams & Wilkins. Retrieved 16 November 2014.

- Engelkirk, Paul; Duben-Engelkirk, Janet (2008). Laboratory Diagnosis of Infectious Diseases. Lippincott Williams & Wilkins. p. 168. Retrieved 16 November 2014.

- Lorian, Victor (2005). Antibiotics in Laboratory Medicine. Lippincott Williams & Wilkins. p. 149. Retrieved 16 November 2014.

- Lee, Mary (2013). Basic Skills in Interpreting Laboratory Data (5 ed.). ASHP. p. 723. Retrieved 16 November 2014.

6

Essential Aspects of Antimicrobial Resistance

The effort to educate people of antimicrobials and to eradicate the overuse of antibiotics is known as antimicrobial stewardship. Some of the essential aspects of antimicrobial resistance discussed in this text are eagle effect, horizontal gene transfer, multidrug tolerance, resistance-nodulation-cell division superfamily, among others. This chapter elucidates the crucial theories and aspects related to antimicrobial resistance.

Antimicrobial Stewardship

Antimicrobial stewardship (AMS) is the systematic effort to educate and persuade prescribers of antimicrobials to follow evidence-based prescribing, in order to stem antibiotic overuse, and thus antimicrobial resistance. AMS has been an organized effort of specialists in infectious diseases,both in Internal Medicine and Pediatrics with their respective peer-organizations, hospital pharmacists, the public health community and their professional organizations since the late 1990s. It has first been implemented in hospitals. In the U.S., AMS has largely been voluntary self-regulation in the form of policies and appeals to adhere to a prescribing discipline. At hospitals, this may take the form of an antimicrobial stewardship program. As of 2014, only the state of California has made AMS mandatory by law.

Definition

AMS, per the 2007 definition of the Society for Healthcare Epidemiology of America (SHEA), is a "set of coordinated strategies to improve the use of antimicrobial medications with the goal to

- enhance patient health outcomes,
- reduce antibiotic resistance, and
- decrease unnecessary costs".

History

Antimicrobial misuse was recognized as early as the 1940s, when Alexander Fleming remarked on penicillin's decreasing efficacy, because of its overuse.

In 1966, the first systematic assessment of antibiotic use in the Winnipeg, Canada general hospital was published: Medical records were reviewed during two non-consecutive four-month periods (medicine, psychiatry, urology, gynecology and surgery, orthopedics, neurosurgery, ear, nose and throat, and ophthalmology). Information was coded on punched cards using 78 columns. Others in 1968 estimated that 50% of antimicrobial use was either unnecessary or inappropriate This figure is likely the lower end of the estimate, and continues to be referenced as of 2015.

In the 1970s the first clinical pharmacy services were established in North American hospitals. The first formal evaluation of antibiotic use in children regarding antibiotic choice, dose and necessity of treatment was undertaken at The Children's Hospital of Winnipeg. Researchers observed errors in therapy in 30% of medical orders and 63% of surgical orders. The most frequent error was unnecessary treatment found in 13% of medical and 45% of surgical orders. The authors stated "Many find it difficult to accept that there are standards against which therapy may be judged."

In the 1980s the antibiotic class of cephalosporins was introduced, further increasing bacterial resistance. During this decade infection control programs began to be established in hospitals, which systematically recorded and investigated hospital-acquired infections. Evidence-based treatment guidelines and regulation of antibiotic use surfaced. Australian researchers published the first medical guideline outcomes research.

The term AMS was coined in 1996 by two internists at Emory University School of Medicine, John McGowan and Dale Gerding, a specialist on C. difficile. They suggested "...large-scale, well-controlled trials of antimicrobial use regulation employing sophisticated epidemiologic methods, molecular biological organism typing, and precise resistance mechanism analysis [...] to determine the best methods to prevent and control this problem [antimicrobial resistance] and ensure our optimal antimicrobial use stewardship" and that "...the long-term effects of antimicrobial selection, dosage, and duration of treatment on resistance development should be a part of every antimicrobial treatment decision."

In 1997, SHEA and the Infectious Diseases Society of America published guidelines to prevent antimicrobial resistance arguing that "...appropriate antimicrobial stewardship, that includes optimal selection, dose, and duration of treatment, as well as control of antibiotic use, will prevent or slow the emergence of resistance among microorganisms."

Ten years later, in 2007, bacterial, antiviral and antifungal resistance had risen to such a degree that the CDC rang the alarm. The same year, IDSA and SHEA published guidelines for developing an AMS program. Also in 2007, the first pediatric publication used the term AMS.

A survey of pediatric infectious disease consultants in 2008 by the Emerging Infectious Disease Network revealed that only 45 (33%) respondents had an AMS program, mostly from before 2000, and another 25 (18%) planned an ASP (data unpublished).

In 2012, the SHEA, IDSA and PIDS published a joint policy statement on AMS.

The CDC's NHSN has been monitoring antimicrobial use and resistance in hospitals that volunteer to provide data.

In 2014, the CDC recommended, that all US hospitals have an antibiotic stewardship program.

The Joint Commission has approved regulations which go into effect January 1, 2017 detailing that hospitals should have an Antimicrobial Stewardship team consisting of Infection preventionist(s),

Pharmacist(s), and a Practitioner to write protocols and develop projects focused on the appropriate use of antibiotics.

Benefits

Decreasing use of antimicrobials has the following proven benefits:

- improves patient outcomes, especially safety
- decreases adverse medication effects such as hypersensitivity reactions, kidney or heart damage (e.g.QT-prolongation).
- decreases Clostridium difficile-associated diarrhea
- decreases invasive candidiasis
- saves costs
- guards the patient's microbiome present in gastrointestinal, respiratory, urogenital tract, and on the skin.
- slows the increase in microbial resistance

Locations

AMS is needed wherever antimicrobials are prescribed in human medicine, namely in hospitals, outpatient clinics, and long term care institutions, including hospice.

Guidelines for prudent or judicious use in veterinary medicine have been developed by the Canadian Veterinary Medicine Association in 2008. A particular problem is that veterinarians are both prescribers and dispensers. Regulators and the veterinary community in the European Union have been discussing the separation of these activities.

Participants

Antimicrobial stewardship focuses on prescribers, be it physician, physician assistant, nurse practitioner, on the prescription and the microorganism, if any. At a hospital, AMS can be organized in the form of an AMS committee that meets monthly. The day-to-day work is done by a core group, usually an infectious disease physician, who may or may not serve in hospital epidemiology and infection control, or/ and a pharmacist, ideally but rarely aided by an information technologist. The entire committee may include physician representatives, who are top antimicrobial prescribers such as physicians in intensive care medicine, Hematology -Oncology, cystic fibrosis clinicians or hospitalists, a microbiologist, a quality improvement (QI) specialist, and a representative from hospital administration.

For an AMS program to be established the institution has to recognize its value. In the US it has become customary to present a business plan to the executive officers of the hospital administration.

Program Components

Thirteen internet-based institutional ASP resources in US academic medical centers have been published. An AMS program has the following tasks, in line with quality improvement theory:

Baseline Assessment

- Measure baseline antimicrobial use, dosing, duration, costs and use patterns.

- Study type of microbial isolates, susceptibilities, and trends thereof

- Identify clinician indications for prescriptions.

In hospitals and clinics using electronic medical records, information technology resources are crucial to home in on these questions. Commercial computer surveillance software programs for microbiology and antimicrobial administrations appear to outnumber "homegrown" institutional programs as of 2015, and include, but are not limited to TREAT Steward, TheraDoc, Sentri7, and Vigilanz.

Goals of Desirable Antimicrobial Use

- Define "appropriate", rational antimicrobial use for the institution, individual patient units, and define empiric treatment versus culture-directed antimicrobial treatment.

- Establish treatment guidelines for clinical syndromes. These can be disseminated in the form of memos, in-services or grand rounds and may be most effective in the form of decision making tools at the point of ordering the prescription.

Interventions on Antimicrobial Prescribing

Provide Feedback, Continuing Education

- Survey prescriber knowledge about antibiotics, antifungal or antiviral drugs.

- Provide targeted education about particular antibiotics, or one specific antimicrobial at a time, as well as empiric treatment for syndromes versus culture directed treatment.

- Assist in making duration more visible to prescribers. Some institutions use automatic stop orders.

- Decreasing diagnostic uncertainty by appropriate testing, including rapid diagnostic methods. The most effective strategy to decrease diagnostic uncertainty would be to align the focus with other safety projects, and QI measures (e.g. blood management, adverse effects etc.).

Biomerieux has published case studies of countries that introduced AMS.

Interventions

The day-to-day work of the core AMS members is to screen patients' medical records for some of the following questions, in order of importance:

- Appropriate antimicrobial choice based on susceptibility, avoiding redundance ?

- Appropriate dose (mg/kg dosing in children) ?

- Appropriate dosing interval according to age, weight and renal function or drug-drug interaction?

- Appropriate deescalation of antimicrobials after culture results are final ?

- Appropriate administration route and feasibility of drug conversion from intravenous to by mouth (PO)?

If the answer is no, the team needs to effectively communicate a recommendation, which may be in person or in the medical record.

Further tasks are:

- Automatic review of the medical record after 72h empiric use, culture results, other laboratory data

- Advise on appropriate duration of antimicrobial therapy

- annual report to administration, calculation of cost savings if any.

Outcomes to Measure

Two pediatric infectious disease physicians have suggested to look at the following variables to judge the outcome of AMS interventions:

- Annual pharmacy acquisition costs

- Antibiotic days/1,000 patient days

- Identifying "drug-bug mismatches"

- IV to oral conversion

- Optimal dosing

- Stopping redundant therapy

- Reducing adverse events

- Overall compliance with ASP recommendations

When examining the relationship between an outcome and an intervention, the epidemiological method of time series analysis is preferred, because it accounts for the dependence between time points. A recent global stewardship survey identified barriers to the initiation, development and implementation of stewardship programmes internationally.

Controversies

- At this time the optimal metrics to benchmark antimicrobial use are still controversial.

- To measure unit of antimicrobials consumed, one can use 'Days Of Therapy' (DOT) or Defined Daily Dose (DDD). The former is more commonly used in the US, the latter is more commonly used in Europe.

- Data source for the use: The electronic medication administration record (eMAR) is the most accurate correlate for doses given, but may be difficult to analyze, because of hold orders and patient refusal, as opposed to administrative or billing data, that may be easier to obtain.

- The question of "appropriateness of use" is probably the most controversial. Appropriate use depends on the local antimicrobial resistance profile and therefore has different regional answers. Merely the "amount" of antibiotics used is no straightforward metric for appropriateness.

- In regard to the most effective AMS intervention, the answer will depend on the size of the institution and the resources available: The system of "prior approval" of antimicrobials by infectious diseases has been used first historically. It is very time- and labor-intensive, and prescribers do not like its restrictive character. Increasingly "post-prescription review" is used.

It can be difficult to decide if a clinical syndrome or a particular drug should be targeted for interventions and education. How to best modify prescriber behavior has been the subject of controversy and research. At issue is how feedback is presented to prescribers, individually, in aggregate, with or without peer comparisons, and whether to reward or punish. As long as the best quality metrics for an AMS program are unknown, a combination of antimicrobial consumption, antimicrobial resistance, and antimicrobial and drug resistant organism related mortality are used.

Eagle Effect

The Eagle effect, Eagle phenomenon, or paradoxical zone phenomenon, named after Harry Eagle who first described it, originally referred to the paradoxically reduced antibacterial effect of penicillin at high doses, though recent usage generally refers to the relative lack of efficacy of beta lactam antibacterial drugs on infections having large numbers of bacteria. The former effect is paradoxical because the effectiveness of an antibiotic generally rises with increasing drug concentration.

Mechanism

Proposed Mechanisms:

- Reduced expression of penicillin binding proteins during stationary growth phase

- Induction of microbial resistance mechanisms (such as beta lactamases with short half-lives) by high drug concentrations

- Precipitation of antimicrobial drug in vitro, possibly also leading to the crystallized drug being mis-detected as colonies of the microbe.

- Self-antagonising the receptor with which it binds (penicillin binding proteins, for example, in the case of a penicillin).

Penicillin is a bactericidal antibiotic that works by inhibiting cell wall synthesis but this synthesis only occurs when bacteria are actively replicating (or in the log phase of growth). In cases of extremely high bacterial burden (such as with Group A Strep), bacteria may be in the stationary phase of growth. In this instance since no bacteria are actively replicating (presumably due to nutrient restriction) penicillin has no activity. This is why adding an antibiotic like clindamycin, which acts ribosomally, kills some of the bacterial and returns them to the log phase of growth.

Horizontal Gene Transfer

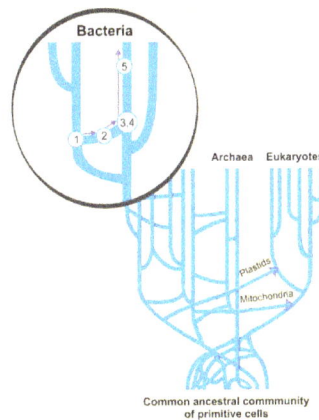

Current tree of life showing vertical and horizontal gene transfers.

Horizontal gene transfer (HGT) is the movement of genetic material between unicellular and/ or multicellular organisms other than via vertical transmission (the transmission of DNA from parent to offspring.) HGT is synonymous with lateral gene transfer (LGT) and the terms are inter-changeable. HGT has been shown to be an important factor in the evolution of many organisms.

Horizontal gene transfer is the primary reason for the spread of antibiotic resistance in bacteria and plays an important role in the evolution of bacteria that can degrade novel compounds such as human-created pesticides and in the evolution, maintenance, and transmission of virulence. This horizontal gene transfer often involves temperate bacteriophages and plasmids. Genes that are responsible for antibiotic resistance in one species of bacteria can be transferred to another species of bacteria through various mechanisms such as F-pilus), subsequently arming the antibiotic resistant genes' recipient against antibiotics, which is becoming a medical challenge to deal with.

Most thinking in genetics has focused upon vertical transfer, but there is a growing awareness that horizontal gene transfer is a highly significant phenomenon and among single-celled organisms, perhaps the dominant form of genetic transfer.

Artificial horizontal gene transfer is a form of genetic engineering.

History

Horizontal genetic transfer was first described in Seattle in 1951 in a publication which demon-strated that the transfer of a viral gene into *Corynebacterium diphtheriae* created a virulent from a non-virulent strain, also simultaneously solving the riddle of diphtheria (that patients could be infected with the bacteria but not have any symptoms, and then suddenly convert later or never), and giving the first example for the relevance of the lysogenic cycle. Inter-bacterial gene transfer was first described in Japan in a 1959 publication that demonstrated the transfer of antibiotic resistance between different species of bacteria. In the mid-1980s, Syvanen predicted that lateral gene transfer existed, had biological significance, and was involved in shaping evolutionary history from the beginning of life on Earth.

As Jian, Rivera and Lake (1999) put it: "Increasingly, studies of genes and genomes are indicating that considerable horizontal transfer has occurred between prokaryotes". The phenomenon appears to have had some significance for unicellular eukaryotes as well. As Bapteste et al. (2005) observe, "additional evidence suggests that gene transfer might also be an important evolutionary mechanism in protist evolution."

There is some evidence that even higher plants and animals have been affected and this has raised concerns for safety. Grafting of one plant to another can transfer chloroplasts (specialised DNA in plants that can conduct photosynthesis), mitichondrial DNA and the entire cell nucleus containing the genome to potentially make a new species. Some Lepidoptera (e.g. Monarch butterflies and silkworms) have been genetically modified by horizontal gene transfer from the wasp bracovirus. Bites from the insect Reduviidae (assassin bug) can, via a parasite, infect humans with the trypanosomal Chagas disease, which can insert its DNA into the human genome. It has been suggested that lateral gene transfer to humans from bacteria may play a role in cancer.

Richardson and Palmer (2007) state: "Horizontal gene transfer (HGT) has played a major role in bacterial evolution and is fairly common in certain unicellular eukaryotes. However, the prevalence and importance of HGT in the evolution of multicellular eukaryotes remain unclear."

Due to the increasing amount of evidence suggesting the importance of these phenomena for evolution molecular biologists such as Peter Gogarten have described horizontal gene transfer as "A New Paradigm for Biology".

Some have argued that the process may be a hidden hazard of genetic engineering as it could allow transgenic DNA to spread from species to species.

Mechanism

There are several mechanisms for horizontal gene transfer:

- Transformation, the genetic alteration of a cell resulting from the introduction, uptake and expression of foreign genetic material (DNA or RNA). This process is relatively common in bacteria, but less so in eukaryotes. Transformation is often used in laboratories to insert novel genes into bacteria for experiments or for industrial or medical applications.

- Transduction, the process in which bacterial DNA is moved from one bacterium to another by a virus (a bacteriophage, or phage).

- Bacterial conjugation, a process that involves the transfer of DNA via a plasmid from a donor cell to a recombinant recipient cell during cell-to-cell contact.

- Gene transfer agents, virus-like elements encoded by the host that are found in the alphaproteobacteria order Rhodobacterales.

A transposon is a mobile segment of DNA that can sometimes pick up a resistance gene and insert it into a plasmid or chromosome, thereby inducing horizontal gene transfer of antibiotic resistance.

Inference

Horizontal gene transfer is typically inferred using bioinformatic methods, either by identifying atypical sequence signatures ("parametric" methods) or by identifying strong discrepancies between the evolutionary history of particular sequences compared to that of their hosts.

Viruses

The virus called *Mimivirus* infects amoebae. Another virus, called *Sputnik*, also infects amoebae, but it cannot reproduce unless mimivirus has already infected the same cell. "Sputnik's genome reveals further insight into its biology. Although 13 of its genes show little similarity to any other known genes, three are closely related to mimivirus and mamavirus genes, perhaps cannibalized by the tiny virus as it packaged up particles sometime in its history. This suggests that the satellite virus could perform horizontal gene transfer between viruses, paralleling the way that bacteriophages ferry genes between bacteria.". Horizontal transfer is also seen between geminiviruses and tobacco plants.

Prokaryotes

Horizontal gene transfer is common among bacteria, even among very distantly related ones. This process is thought to be a significant cause of increased drug resistance when one bacterial cell acquires resistance, and the resistance genes are transferred to other species. Transposition and horizontal gene transfer, along with strong natural selective forces have led to multi-drug resistant strains of *S. aureus* and many other pathogenic bacteria. Horizontal gene transfer also plays a role in the spread of virulence factors, such as exotoxins and exoenzymes, amongst bacteria. A prime example concerning the spread of exotoxins is the adaptive evolution of Shiga toxins in *E. coli* through horizontal gene transfer via transduction with *Shigella* species of bacteria. Strategies to combat certain bacterial infections by targeting these specific virulence factors and mobile genetic elements have been proposed. For example, horizontally transferred genetic elements play important roles in the virulence of E. coli, Salmonella, Streptococcus and Clostridium perfringens.

In prokaryotes, restriction-modification systems are known to provide immunity against horizontal gene transfer and in stabilizing mobile genetic elements. Genes encoding restriction modification systems systems have been reported to move between prokaryotic genomes within mobile genetic elements such as plasmids, prophages, insertion sequences/transposons, integrative conjugative elements (ICEs) and integrons. Still, they are more frequently a chromosomal-encoded barrier to MGEs than an MGE-encoded tool for cell infection.

Eubacterial Transformation

Natural transformation is a bacterial adaptation for DNA transfer (HGT) that depends on the expression of numerous bacterial genes whose products are responsible for this process. In general, transformation is a complex, energy-requiring developmental process. In order for a bacterium to bind, take up and recombine exogenous DNA into its chromosome, it must become competent, that is, enter a special physiological state. Competence development in Bacillus subtilis requires expression of about 40 genes. The DNA integrated into the host chromosome is usually (but with infrequent exceptions) derived from another bacterium of the same species, and is thus homologous to the resident chromosome. The capacity for natural transformation occurs in at least 67 prokaryotic species. Competence for transformation is typically induced by high cell density and/ or nutritional limitation, conditions associated with the stationary phase of bacterial growth. Competence appears to be an adaptation for DNA repair. Transformation in bacteria can be viewed as a primitive sexual process, since it involves interaction of homologous DNA from two individuals to form recombinant DNA that is passed on to succeeding generations.

Eubacterial Conjugation

Conjugation in *Mycobacterium smegmatis*, like conjugation in *E. coli*, requires stable and extended contact between a donor and a recipient strain, is DNase resistant, and the transferred DNA is incorporated into the recipient chromosome by homologous recombination. However, unlike *E. coli* high frequency of recombination conjugation (Hfr), mycobacterial conjugation is a type of HGT that is chromosome rather than plasmid based. Furthermore, in contrast to *E. coli* (Hfr) conjugation, in M. smegmatis all regions of the chromosome are transferred with comparable efficiencies. Substantial blending of the parental genomes was found as a result of conjugation, and this blending was regarded as reminiscent of that seen in the meiotic products of sexual reproduction.

Archaeal DNA Transfer

The archaeon *Sulfolobus solfataricus*, when UV irradiated, strongly induces the formation of type IV pili which then facilitates cellular aggregation. Exposure to chemical agents that cause DNA damage also induces cellular aggregation. Other physical stressors, such as temperature shift or pH, do not induce aggregation, suggesting that DNA damage is a specific inducer of cellular aggregation.

UV-induced cellular aggregation mediates intercellular chromosomal HGT marker exchange with high frequency, and UV-induced cultures display recombination rates that exceed those of uninduced cultures by as much as three orders of magnitude. *S. solfataricus* cells aggregate preferentially with other cells of their own species. Frols et al. and Ajon et al. suggested that UV-inducible DNA transfer is likely an important mechanism for providing increased repair of damaged DNA via homologous recombination. This process can be regarded as a simple form of sexual interaction.

Another thermophilic species, *Sulfolobus acidocaldarius*, is able to undergo HGT. *S. acidocaldarius* can exchange and recombine chromosomal markers at temperatures up to 84°C. UV exposure induces pili formation and cellular aggregation. Cells with the ability to aggregate have greater survival than mutants lacking pili that are unable to aggregate. The frequency of recombination is increased by DNA damage induced by UV-irradiation and by DNA damaging chemicals.

The *ups* operon, containing five genes, is highly induced by UV irradiation. The proteins encoded by the *ups* operon are employed in UV-induced pili assembly and cellular aggregation leading to intercellular DNA exchange and homologous recombination. Since this system increases the fitness of *S. acidocaldarius* cells after UV exposure, Wolferen et al. considered that transfer of DNA likely takes place in order to repair UV-induced DNA damages by homologous recombination.

Eukaryotes

"Sequence comparisons suggest recent horizontal transfer of many genes among diverse species including across the boundaries of phylogenetic 'domains'. Thus determining the phylogenetic history of a species can not be done conclusively by determining evolutionary trees for single genes".

- Analysis of DNA sequences suggests that horizontal gene transfer has occurred within eukaryotes from the chloroplast and mitochondrial genomes to the nuclear genome. As stated in the endosymbiotic theory, chloroplasts and mitochondria probably originated as bacterial endosymbionts of a progenitor to the eukaryotic cell.

- Horizontal transfer occurs from bacteria to some fungi, such as the yeast *Saccharomyces cerevisiae*.

- The adzuki bean beetle has acquired genetic material from its (non-beneficial) endosymbiont *Wolbachia*. New examples have recently been reported demonstrating that Wolbachia bacteria represent an important potential source of genetic material in arthropods and filarial nematodes.

- Mitochondrial genes moved to parasites of the Rafflesiaceae plant family from their hosts and from chloroplasts of a not-yet-identified plant to the mitochondria of the bean *Phaseolus*.

- *Striga hermonthica*, a eudicot, has received a gene from sorghum (*Sorghum bicolor*) to its nuclear genome. The gene is of unknown functionality.

- Pea aphids (*Acyrthosiphon pisum*) contain multiple genes from fungi. Plants, fungi and microorganisms can synthesize carotenoids, but torulene made by pea aphids is the only carotenoid known to be synthesized by an organism in the animal kingdom.

- The malaria pathogen *Plasmodium vivax* acquired genetic material from humans that might help facilitate its long stay in the body.

- A bacteriophage-mediated mechanism transfers genes between prokaryotes and eukaryotes. Nuclear localization signals in bacteriophage terminal proteins (TP) prime DNA replication and become covalently linked to the viral genome. The role of virus and bacteriophages in HGT in bacteria, suggests that TP-containing genomes could be a vehicle of inter-kingdom genetic information transference all throughout evolution.

- HhMAN1 is a gene in the genome of the coffee borer beetle (*Hypothenemus hampei*) that resembles bacterial genes, and is thought to be transferred from bacteria in the beetle's gut.

- A gene that allowed ferns to survive in dark forests came from the hornwort, which grows in mats on streambanks or trees. The neochrome gene arrived about 180 million years ago.

- Plants are capable of receiving genetic information from viruses by horizontal gene transfer.

- One study identified approximately 100 of humans' approximately 20,000 total genes which likely resulted from horizontal gene transfer, but this number has been challenged by several researchers arguing these candidate genes for HGT are more likely the result of gene loss combined with differences in the rate of evolution

- Bdelloid rotifers currently hold the 'record' for HGT in animals with ~8% of their genes from bacterial origins. Tardigrades were thought to break the record with 17.5% HGT, but that finding was an artifact of bacterial contamination.

- A study found the genomes of 40 animals (including 10 primates, four Caenorhabditis worms and 12 Drosophila insects) contained genes which the researchers concluded had been transferred from bacteria and fungi by horizontal gene transfer. The researchers estimated that for some nematodes and Drosophilia insects these genes had been acquired relatively recently.

Horizontal Transposon Transfer (HTT)

Horizontal transposon transfer (HTT) refers to the passage of pieces of DNA that are characterized by their ability to move from one locus to another between genomes by means other than parent-to-offspring inheritance. Horizontal gene transfer has long been thought to be crucial to prokaryotic evolution, but there is a growing amount of data showing that HTT is a common and widespread phenomenon in eukaryote evolution as well (). On the transposable element (TE) side, spreading between genomes via horizontal transfer may be viewed as a strategy to escape purging due to purifying selection, mutational decay and/or host defense mechanisms () .

HTT can occur with any type of transposable elements, but DNA transposons and LTR retroelements are more likely to be capable of HTT because both have a stable, double-stranded DNA intermediate that is thought to be sturdier than the single-stranded RNA intermediate of non-LTR retroelements, which can be highly degradable (). Non-autonomous elements may be less likely to transfer horizontally compared to autonomous elements because they do not encode the proteins required for their own mobilization. The structure of these non-autonomous elements generally consists of an intronless gene encoding a transposase protein, and may or may not have a promoter sequence. Those that do not have promoter sequences encoded within the mobile region rely on adjacent host promoters for expression (). Horizontal transfer is thought to play an important role in the TE lifecycle ().

HTT has been shown to occur between species and across continents in both plants () and animals (Ivancevic et al. 2013), though some TEs have been shown to more successfully colonize the genomes of certain species over others (). Both spatial and taxonomic proximity of species has been proposed to favor HTTs in plants and animals (). It is unknown how the density of a population may affect the rate of HTT events within a population, but close proximity due to parasitism and cross contamination due to crowding have been proposed to favor HTT in both plants and animals (). Successful transfer of a transposable element requires delivery of DNA from donor to host cell (and to the germ line for multi-cellular organisms), followed by integration into the recipient host genome (). Though the actual mechanism for the transportation of TEs from donor cells to host cells is unknown, it is established that naked DNA and RNA can circulate in bodily fluid (). Many proposed vectors include arthropods, viruses, freshwater snails (Ivancevic et al. 2013), endosymbiotic bacteria (), and intracellular parasitic bacteria (). In some cases, even TEs facilitate transport for other TEs ().

The arrival of a new TE in a host genome can have detrimental consequences because TE mobility may induce mutation. However, HTT can also be beneficial by introducing new genetic material into a genome and promoting the shuffling of genes and TE domains among hosts, which can be co-opted by the host genome to perform new functions (). Moreover, transposition activity increases the TE copy number and generates chromosomal rearrangement hotspots (). HTT detection is a difficult task because it is an ongoing phenomenon that is constantly changing in frequency of occurrence and composition of TEs inside host genomes. Furthermore, few species have been analyzed for HTT, making it difficult to establish patterns of HTT events between species. These issues can lead to the underestimation or overestimation of HTT events between ancestral and current eukaryotic species ().

Artificial Horizontal Gene Transfer

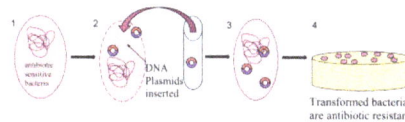

Before it is transformed a bacterium is susceptible to antibiotics. A plasmid can be inserted when the bacteria is under stress, and be incorporated into the bacterial DNA creating antibiotic resistance. When the plasmids are prepared they are inserted into the bacterial cell by either making pores in the plasma membrane with temperature extremes and chemical treatments, or making it semi permeable through the process of electrophoresis, in which electric currents create the holes in the membrane. After conditions return to normal the holes in the membrane close and the plasmids are trapped inside the bacteria where they become part of the genetic material and their genes are expressed by the bacteria.

Genetic engineering is essentially horizontal gene transfer, albeit with synthetic expression cassettes. The Sleeping Beauty transposon system (SB) was developed as a synthetic gene transfer agent that was based on the known abilities of Tc1/mariner transposons to invade genomes of extremely diverse species. The SB system has been used to introduce genetic sequences into a wide variety of animal genomes.

Importance in Evolution

Horizontal gene transfer is a potential confounding factor in inferring phylogenetic trees based on the sequence of one gene. For example, given two distantly related bacteria that have exchanged a gene a phylogenetic tree including those species will show them to be closely related because that gene is the same even though most other genes are dissimilar. For this reason it is often ideal to use other information to infer robust phylogenies such as the presence or absence of genes or, more commonly, to include as wide a range of genes for phylogenetic analysis as possible.

For example, the most common gene to be used for constructing phylogenetic relationships in prokaryotes is the 16s rRNA gene since its sequences tend to be conserved among members with close phylogenetic distances, but variable enough that differences can be measured. However, in recent years it has also been argued that 16s rRNA genes can also be horizontally transferred. Although this may be infrequent, the validity of 16s rRNA-constructed phylogenetic trees must be reevaluated.

Biologist Johann Peter Gogarten suggests "the original metaphor of a tree no longer fits the data from recent genome research" therefore "biologists should use the metaphor of a mosaic to describe the different histories combined in individual genomes and use the metaphor of a net to visualize the rich exchange and cooperative effects of HGT among microbes." There exist several methods to infer such phylogenetic networks.

Using single genes as phylogenetic markers, it is difficult to trace organismal phylogeny in the presence of horizontal gene transfer. Combining the simple coalescence model of cladogenesis with rare HGT horizontal gene transfer events suggest there was no single most recent common ancestor that contained all of the genes ancestral to those shared among the three domains of life. Each contemporary molecule has its own history and traces back to an individual molecule cenancestor. However, these molecular ancestors were likely to be present in different organisms at different times."

Scientific American Article (2000)

Uprooting the Tree of Life by W. Ford Doolittle (*Scientific American*, February 2000, pp 90–95) contains a discussion of the Last Universal Common Ancestor and the problems that arose with respect to that concept when one considers horizontal gene transfer. The article covers a wide area — the endosymbiont hypothesis for eukaryotes, the use of small subunit ribosomal RNA (SSU rRNA) as a measure of evolutionary distances (this was the field Carl Woese worked in when formulating the first modern "tree of life", and his research results with SSU rRNA led him to propose the Archaea as a third domain of life) and other relevant topics. Indeed, it was while examining the new three-domain view of life that horizontal gene transfer arose as a complicating issue: *Archaeoglobus fulgidus* is cited in the article (p. 76) as being an anomaly with respect to a phylogenetic tree based upon the encoding for the enzyme HMGCoA reductase — the organism in question is a definite Archaean, with all the cell lipids and transcription machinery that are expected of an Archaean, but whose HMGCoA genes are actually of bacterial origin.

Again on p. 76, the article continues with:

> "The weight of evidence still supports the likelihood that mitochondria in eukaryotes derived from alpha-proteobacterial cells and that chloroplasts came from ingested cyanobacteria, but it is no longer safe to assume that those were the only lateral gene transfers that occurred after the first eukaryotes arose. Only in later, multicellular eukaryotes do we know of definite restrictions on horizontal gene exchange, such as the advent of separated (and protected) germ cells."

The article continues with:

> "If there had never been any lateral gene transfer, all these individual gene trees would have the same topology (the same branching order), and the ancestral genes at the root of each tree would have all been present in the last universal common ancestor, a single ancient cell. But extensive transfer means that neither is the case: gene trees will differ (although many will have regions of similar topology) *and* there would never have been a single cell that could be called the last universal common ancestor.

> "As Woese has written, 'the ancestor cannot have been a particular organism, a single organismal lineage. It was communal, a loosely knit, diverse conglomeration of primitive cells that evolved as a unit, and it eventually developed to a stage where it broke into several distinct communities, which in their turn became the three primary lines of descent (bacteria, archaea and eukaryotes)' In other words, early cells, each having relatively few genes, differed in many ways. By swapping genes freely, they shared various of their tal-

ents with their contemporaries. Eventually this collection of eclectic and changeable cells coalesced into the three basic domains known today. These domains become recognisable because much (though by no means all) of the gene transfer that occurs these days goes on within domains."

With regard to how horizontal gene transfer affects evolutionary theory (common descent, universal phylogenetic tree) Carl Woese says:

"What elevated common descent to doctrinal status almost certainly was the much later discovery of the universality of biochemistry, which was seemingly impossible to explain otherwise. But that was before horizontal gene transfer (HGT), which could offer an alternative explanation for the universality of biochemistry, was recognized as a major part of the evolutionary dynamic. In questioning the doctrine of common descent, one necessarily questions the universal phylogenetic tree. That compelling tree image resides deep in our representation of biology. But the tree is no more than a graphical device; it is not some a priori form that nature imposes upon the evolutionary process. It is not a matter of whether your data are consistent with a tree, but whether tree topology is a useful way to represent your data. Ordinarily it is, of course, but the universal tree is no ordinary tree, and its root no ordinary root. Under conditions of extreme HGT, there is no (organismal) "tree." Evolution is basically reticulate."

In a May 2010 article in *Nature*, Douglas Theobald argued that there was indeed one Last Universal Common Ancestor to all existing life and that horizontal gene transfer has not destroyed our ability to infer this.

Genes

This list is incomplete; you can help by expanding it.

There is evidence for historical horizontal transfer of the following genes:

- Lycopene cyclase for carotenoid biosynthesis, between Chlorobi and Cyanobacteria.

- *TetO* gen conferring resistance to tetracycline, between *Campylobacter jejuni*.

- Neochrome, gene in some ferns that enhances their ability to survive in dim light. Believed to have been acquired from algae sometime during the Cretaceous.

- transfer of a cysteine synthase from a bacteria into phytophagous mites and Lepidoptera allowing the detoxification of cyanogenic glucosides produced by host plants.

- The LINE1 sequence has transferred from humans to the gonorrhea bacteria.

Multidrug Tolerance

Multidrug tolerance or antibiotic tolerance is the ability of a disease-causing microorganism to resist killing by antibiotics or other antimicrobials. It is mechanistically distinct from multidrug resistance: It is not caused by mutant microbes, but rather by microbial cells that exist in a transient, dormant, non-dividing state. Microorganisms that display multidrug tolerance can be bacteria, fungi or parasites.

History

Recognition of antibiotic tolerance dates back to 1944 when Joseph Bigger, an Irish physician working in England, was experimenting with the recently discovered penicillin. Bigger used penicillin to lyse a suspension of bacteria and then inoculated culture medium with the penicillin-treated liquid. Colonies of bacteria were able to grow after antibiotic killing. The important observation that Bigger made was that this new population could be again be killed by penicillin except for a small residual population. Hence the residual organisms were not antibiotic resistant mutants but rather a subpopulation of what he called 'persisters'. The formation of persisters is now known to be a common phenomenon that can occur in response to a variety of antibiotics.

Relevance to Chronic Infections

Multidrug tolerance is caused by a small subpopulation of microbial cells termed persisters. Persisters are not mutants, but rather are dormant cells that can survive the antimicrobial treatments that kill the majority of their genetically identical siblings. Persister cells have entered a non- or extremely slow-growing physiological state which makes them insensitive (refractory or tolerant) to the action of antimicrobial drugs. When such persisting microbial cells cannot be eliminated by the immune system, they become a reservoir from which recurrence of infection will develop. Indeed, it appears that persister cells are the main cause for relapsing and chronic infections. Chronic infections can affect people of any age, health, or immune status.

Medical importance

Bacterial multidrug or antibiotic tolerance poses medically important challenges. It is largely responsible for the inability to eradicate bacterial infections with antibiotic treatment. Persister cells are highly enriched in biofilms, and it has been suggested that this is the reason that makes biofilm-related diseases so hard to treat. Examples are chronic infections of implanted medical devices such as catheters and artificial joints, urinary tract infections, middle ear infections and fatal lung disease .

Distinction from Multidrug Resistance

Unlike resistance, multidrug tolerance is a transient, non-heritable phenotype. Multidrug tolerant persister cells are not antibiotic resistant mutants. Resistance is caused by newly acquired genetic traits (by mutation or horizontal gene transfer) that are heritable and confer the ability to grow at elevated concentrations of antimicrobial drugs. In contrast, multidrug tolerance is caused by a reversible physiological state in a small subpopulation of genetically identical cells, similar to a differentiated cell type. It enables this small subpopulation of microbes to survive the antibiotic killing of their surrounding siblings. Persisting cells resume growth when the antimicrobial agent is removed, and their progeny is sensitive to antimicrobial agents.

Molecular Mechanisms

The molecular mechanisms that underlie persister cell formation and multidrug tolerance are largely unknown. Persister cells are thought to arise spontaneously in a growing microbial population by a stochastic genetic switch, although inducible mechanisms of persister cell formation have been described. Owing to their transient nature and relatively low abundance, it is hard to isolate persister cells in sufficient numbers for experimental characterization, and only a few relevant genes have been identified to date. The best-understood persistence factor is the *E. coli* *hi*gh persistence gene, commonly abbreviated as *hipA*.

Although tolerance is widely considered a passive state, there is evidence indicating it can be an energy-dependent process. Persister cells in *E. coli* can transport intracellular accumulations antibiotic using an energy requiring efflux pump called TolC.

Potential Treatment

In May 2011, it was reported by Nature.com that the addition of certain metabolites can help suppress multidrug tolerance in numerous species of bacteria, including *E. coli* and *S. aureus*, by "the generation of a proton-motive force which facilitates aminoglycoside uptake". Phage therapy, where applicable, entirely circumvents antibiotic resistance.

Resistance-nodulation-cell Division Superfamily

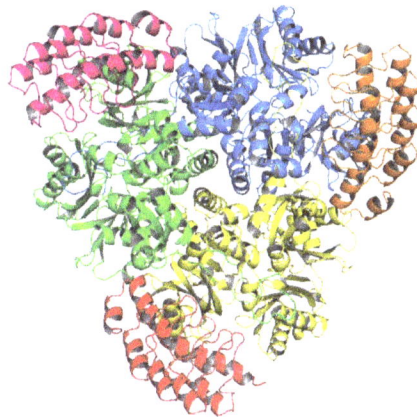

Crystallized AcrB: An HAE-RND subclass protein involved in drug and amphiphilic efflux

Resistance-nodulation-division (RND) family transporters are a category of bacterial efflux pumps, especially identified in Gram-negative bacteria and located in the cytoplasmic membrane, that actively transport substrates. The RND superfamily includes seven families: the heavy metal efflux (HME), the hydrophobe/amphiphile efflux-1 (gram-negative bacteria), the nodulation factor exporter family (NFE), the SecDF protein-secretion accessory protein family, the hydrophobe/amphiphile efflux-2 family, the eukaryotic sterol homeostasis family, and the hydrophobe/amphiphile efflux-3 family. These RND systems are involved in maintaining homeostasis of the cell, removal of toxic compounds, and export of virulence determinants. They have a broad substrate spectrum and can lead to the diminished activity of unrelated drug classes if over-expressed. The first reports of drug resistant bacterial infections were reported in the 1940s after the first mass production of antibiotics. Most of the RND superfamily transport systems are made of large polypeptide chains. RND proteins exist primarily in gram-negative bacteria but can also be found in gram-positive bacteria, archaea, and eukaryotes.

Function

The RND protein dictates the substrate for the completed transport systems including: metal ions, xenobiotics or drugs. Transport of hydrophobic and amphiphilic compounds are carried out by the HAE-RND subfamily. While the efflux of heavy metals are preformed HME-RND.

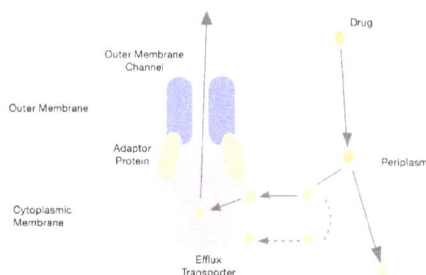

Triparitate Complex Model: RND inner-membrane protein, outer-membrane fusion protein, & periplasmic adaptor protein.

Mechanism and Structure

RND proteins are large and can include more than 1000 amino acid residues. They are generally composed of two homologous subunits (suggesting they arose as a result of an intragenic tandem duplication event that occurred in the primordial system prior to divergence of the family members) each containing a periplasmic loop adjacent to 12 transmembrane helices. Of the twelve helices there is a single transmembrane spanner (TMS) at the N-terminus followed by a large extracytoplasmic domain, then six additional TMSs, a second large extracytoplasmic domain, and five final C-terminal TMSs. TM4 governs the specificity for a particular substrate in a given RND protein. Therefore, TM4 can be an indicator for RND specificity without explicit knowledge of the remainder of the protein.

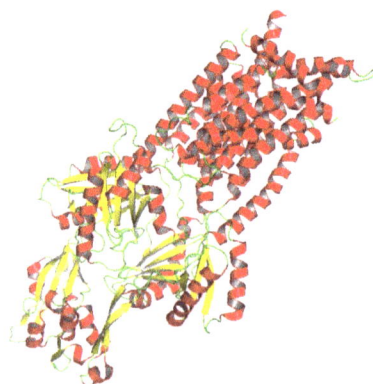

Crystallized CusA: HAE-RND subclass protein

RND pumps are the cytoplasmic residing portion of a complete tripartite complex (Fig. 1) which spreads across the outer-membrane and the inner membrane of gram-negative bacteria, also commonly referred to as the CBA efflux system. The RND protein associates with an outer membrane channel and a periplasmic adaptor protein, and the association of all three proteins allows the system to export substrates into the external medium, providing a huge advantage for the bacteria.

The CusA protein, a HME-RND member transporter, was able to be crystallized providing valuable structural information of HME-RND pumps. CusA exists as a homotrimer with each unit consisting of 12 transmembrane helices (TM1-TM12). The periplasmic domain consists of two helices, TM2 and TM8. In addition, the periplasmic domain is made up of six subdomains, PN1, PN2, PC1, PC2, DN, DC, which form a central poor and a dock domain. The central pore is formed by PN1, PN2, PC1, PC2, and together stabilize the trimeric organization of the homotrimer.

Metal Ion Efflux (HME-RND)

The HME-RND family functions as the central protein pump in metal ion efflux powered by a proton-substrate antiport. The family includes pumps which export monovalent metals—the Cus system, and pumps which export diavlent metals—the Czc system.

Heavy metal resistance by the RND family was first discovered in *R. metallidurans* through the CzcA and later the CnrA protein. The best characterized RND proteins include CzcCBA (Cd^{2+}, Zn^{2+}, and Co^{2+}), CnrCBA (Ni^{2+} and Co^{2+}), and NccCBA (Ni^{2+}, Co^{2+} and Cd^{2+}) in *Cupriavidus,* Czr (Cd^{2+} and Zn^{2+} resistance) in *Pseudomonas aeruginosa,* and Czn (Cd^{2+}, Zn^{2+}, and Ni^{2+} resistance) in *Helicobacter pylori.* It has been proposed that metal-ion efflux occurs from the cytoplasm and periplasm based on the location of multiple substrate binding sites on the RND protein.

CznCBA

The Czn system maintains homeostasis of Cadmium, Zinc, and Nickel resistance; it is involved in Urease modulation, and gastric colonization by *H. pylori*. The CznC and CznA proteins play the dominating role in nickel homeostasis.

CzcCBA

Czc confers resistance to Cobalt, Zinc, and Cadmium. The CzcCBA operon includes: CzcA (the RND family specific protein), the membrane fusion protein (MFP) CzcB, and the outer membrane factor protein (OMF) CzcC, all of which form the active tripartite complex, and the *czcoperon*. Expression of the operon is regulated through metal ions.

Drug Resistance (HAE-RND)

The RND family plays an important role in producing intrinsic and elevated multi-drug resistance in gram-negative bacteria. The export of amphiphilic and hydrophobic substrates is governed by the HAE-RND family. In *E. coli* five RND pumps have been specifically identified: AcrAB, AcrAD, AcrEF, MdtEF, and MdtAB. Although it is not clear how the tripartite complex works in bacteria two mechanisms have been proposed: *Adaptor Bridging Model* and *Adaptor Wrapping Model*.

HAE-RNDs involvement in the detoxification and exportation of organic substrates allowed for recent characterization of specific pumps due to their increasing medical relevance. Half of the antibiotic resistance demonstrated in *in vivo* hospital strains of *Pseduomonas aeruginosa* was attributed to RND efflux proteins. *P. aeruginosa* contain 13 RND transport systems, including one HME-RND and the remaining HAE-RNDs. Among the best identified are the Mex proteins: MexB, MexD, and MexF, which detoxify organic substances. It is proposed that the MexB systems demonstrates substrate specificity for beta-lactams; while the MexD-system expresses specificity for cepheme comounds.

E. coli - AcrB

In *E. coli* multi-drug resistance develops from a variety of mechanisms. Particularly concerning is the ability of efflux mechanisms to confer broad band resistivity. RND efflux pumps provide extrusion for a range of compounds. Five protein transporters in *E. coli* cells that belong to the HAE-RND subfamily have been classified, including the multi-drug efflux protein AcrB, the outer membrane protein TolC and the periplasmic adaptor protein AcrA. The TolC and AcrA proteins are also utilized in the tripartite complex in other identified RND efflux proteins. The AcrAB-TolC efflux system is responsible for the efflux of antimicrobial drugs like penicillin G, cloxacillin, nafcillin, macrolides, novobiocin, linezolid, and fusidic acid antibiotics. Other substrates include dyes, detergents, some organic solvents, and steroid hormones. The ways in which the lipophilic domains of the substrate and the RND pumps is not completely defined.

The crystallized AcrB protein, provides insight into the mechanism of action of HAE-RND proteins, and other RND family proteins.

Multidrug Transport (Mdt) Efflux

Mdt(A) is an efflux pump that confers resistance to a variety of drugs. It is expressed in *L. lactis, E. coli* and various other bacteria. Unlike other RND proteins Mdt(A) contains a putative ATP-binding site and two C-motifs conserved in its fifth TMS. Mdt is effective at providing the bacteria with resistance to tetracycline, chloramphenicol, lincosamides and streptomycin. The source of energy for active efflux by Mdt(A) is currently unknown.

References

- Pollard, Andrew J.; McCracken, George H.; Finn, Adam (2004). Hot Topics in Infection and Immunity in Children. Springer. p. 187. ISBN 9780306483448.

- El Baidouri Moaine; et al. (2014). "Widespread and Frequent Horizontal Transfers of Transposable Elements in Plants". Genome Research. 24 (5): 831–838. doi:10.1101/gr.164400.113.

- Darkened Forests, Ferns Stole Gene From an Unlikely Source — and Then From Each Other by Jennifer Frazer (May 6, 2014). Scientific American.

- Oliveira, PH; Touchon, M; Rocha, EPC (2014). "The interplay of restriction-modification systems with mobile genetic elements and their prokaryotic hosts". Nucleic Acids Res. 42 (16): 10618–10631. doi:10.1093/nar/gku734.

- Morris, Brener S, S (2012). "Use of a structured panel process to define quality metrics for antimicrobial stewardship programs.". Infect Ctrl Hosp Epidemiol. 33: 500–6. doi:10.1086/665324. PMID 22476277. Retrieved 7 March 2014.

- HICPAC (March 2013). "Updates on NHSN Monitoring of Antimicrobial Use and Resistance" (PDF). Retrieved 1 June 2014.

- Shlaes, D; et al. (April 1997). "Guidelines for the prevention of antimicrobial resistance in hospitals." Check |url= value (help). Infect Control Hosp Epidemiol. 418: 275–91. PMID 9131374. Retrieved 7 March 2014.

- Francino, MP (editor) (2012). Horizontal Gene Transfer in Microorganisms. Caister Academic Press. ISBN 978-1-908230-10-2.

- Le Page, Michael (2016-03-17). "Farmers may have been accidentally making GMOs for millennia". The New Scientist. Retrieved 2016-07-11.

- Yong, Ed (2010-02-14). "Genes from Chagas parasite can transfer to humans and be passed on to children". National Geographic. Retrieved 2016-07-13.

- Madhusoodanan, Jyoti (2015-03-12). "Horizontal Gene Transfer a Hallmark of Animal Genomes?". The Scientist. Retrieved 2016-07-14.

- Dupeyron, Mathilde et al. "Horizontal Transfer of Transposons between and within Crustaceans and Insects." Mobile DNA 5 (2014): 4. PMC. Web. 4 Mar. 2016.

- Wallau, Gabriel Luz, Mauro Freitas Ortiz, and Elgion Lucio Silva Loreto. "Horizontal Transposon Transfer in Eukarya: Detection, Bias, and Perspectives." Genome Biology and Evolution 4.8 (2012): 801–811. PMC. Web. 4 Mar. 2016.

- Yong, Ed (2011-02-16). "Gonorrhea has picked up human DNA (and that's just the beginning)". National Geographic. Retrieved 2016-07-14.

- Pu, Y, et al. (2016) Enhanced Efflux Activity Facilitates Drug Tolerance in Dormant Bacterial Cells. Mol Cell, 62: 284–294.

- "Human beings' ancestors have routinely stolen genes from other species". The Economist. 14 March 2015. Retrieved 17 March 2015.

- Anes, João, et al. "The ins and outs of RND efflux pumps in Escherichia coli." Frontiers in microbiology 6 (2015). PMID 26113845

Permissions

Index

www.ingramcontent.com/pod-product-compliance
Lightning Source LLC
Chambersburg PA
CBHW061252190326

41458CB00011B/3648